MANAGEMENT PLANNING

MANAGEMENT PLANNING

A Systems Approach

Norbert Lloyd Enrick, Ph. D.

Professor of Management and Production
Kent State University

McGRAW-HILL BOOK COMPANY

NEW YORK   SAN FRANCISCO   TORONTO   LONDON   SYDNEY

MANAGEMENT PLANNING

19524

1234567890 MAMM 75432106987

FOREWORD

Efficient business and industrial performance depends upon well-laid and smoothly functioning plans and operations. Productivity, quality, sales volume, and overall profitability all are among the measures of effective management planning. Without coordination of plans, schedules, and goals we invite waste and inefficiency and the frustrations that emanate from working at cross-purposes and similar nonplanning.

In our age of innovation, spurred by the rapidly expanding pace of technology, with the concomitant changes in production and marketing, it is becoming increasingly difficult to maintain coordination of all phases of a business organization. Changes are swift, competition keen. Increasingly we learn that it is no longer acceptable to develop plans, schedules, and programs that will merely "work" or even tend to work smoothly without major hitch. *Planning must be optimal.* Among the many alternatives of choice as to what new equipment to purchase, what new products to market, and how to mesh sales potentials with productive capacities, management is called on to *choose those paths and only those that will be best in terms of ultimate profitability.*

But as technology advances, products proliferate, and new markets continue to open up, the choice of alternatives and their interlocking combinations of production, engineering, sales, and promotion rises exponentially. Which machinery will be best suited for our needs? What products should we stress in the coming season's promotions? How much stock should we build up for each model and style? These all are interdependent questions, for the promotional plans may call for certain types of new products and machinery to make it, but the costs of acquiring the new equipment and producing the new items may force a judicious alteration of the initial promotion plans. Factors of quality, price structures, competitive strategies, and long-range vs short-range trends all play their further part in the complex decision structure. The human mind may well despair at having to juggle all of the pertinent figures and cost and other relationships in arriving at the optimal path.

In response, the modern quantitatively oriented methods of Management Science have been called in as aids to the executive in his planning, decision-making, and control work.

Among the quantitative techniques of Management Science, one of the key methods of probably the widest usefulness is Mathematical Programming, MP. Management searches out and identifies the principal interlocking factors of a decision-problem. Information is quantified to the extent possible. Interrelations are expressed in the form of simple mathematical equations. These equations thus represent a mathematical model of the real world, such as the production rates for various items in relation to the available capacities limiting production, profits as a function of costs and quantity in comparison to selling price and the limits of saleability, and other such relations. Next, an MP analysis of the model—usually performed on a computer—comes up with results giving management a deep and penetrating insight into the structure of the relationships involved and their effects on costs, profitability, or other measures of operational effectiveness. Aided by this information, management will be in a better position to shape policies, lay plans, adopt programs, set up schedules, and arrive at decisions that are truly optimum-directed. Time and again, when an MP analysis has been made in an organization, it is found that the program thus recommended may be superior to the prior accomplishments of unaided intuitive judgment. As a general rule, it can be said that MP helps management in the decision-making process, so that in the aggregate a 20% improvement in profit over nonmathematical planning may be considered a realistically expected gain.

MP serves not only in the executive suite but throughout the organization, to the foreman scheduling his machines, the designer aiming to engineer quality and reliability within given cost limitations into the new product, and the production manager planning required inventory buildups. In distribution and sales, similarly, MP can aid both in the overall strategy planning and in within-territory scheduling. References are also made in this book to the extension of MP techniques to other management areas, such as the planning of military operations for maximum effectiveness at a relative minimum of losses (Military Operations Research) or in developing public service programs, such as for police protection or highway networks (Public Science).

Programming does not imply a rigid straightjacket. Rather it refers to the fact that the MP analysis provides a recommended program for management's consideration. If, in the light of managerial judgment, this program is workable and likely to lead to the desired end results, then it will be adopted. Management judgment is critically needed not only in identifying the problem data and variables that go into the original program analysis, but also in evaluating the outcomes of the mathematical procedures and in making such modifications and revisions of the final program as will be needed from a practical standpoint.

Though MP will be highly mathematical in the computer stage that leads to the recommended program, let it be stated emphatically that you *need not be a mathematician* either to read this book or to take full advantage of this Management Science tool. But you must be capable of properly assessing and adequately discerning within your managerial environment:

— Where is an MP analysis needed and likely to be of value?
— What types of data will have to be gathered?
— Which interlocking functions must be analyzed?
— How to interpret the results.
— How to utilize and apply the MP solution with such modifications and adaptations as the practical situation demands.

Admittedly we did provide a mathematical chapter near the end of this book, with the primary intent to serve the student who is interested in this. The managerial content of this chapter is negligible, and it may thus be safely skipped by the practitioner, who may not have time for the theoretical background. Similarly practice problems are given. These are not mandatory, but trying your hand at them may be most helpful in developing skills at recognizing situations in which MP can be of value.

The hope is, thus, that this book will serve the practical manager and the student alike by bringing a clear, use–oriented introduction to the quantitative Management Science methods of Mathematical Programming for operations planning. A reading should enable you not only to understand the basic principles, methods, and philosophy involved, but also to be effective either as an originator of MP installations in an organization or as a working member of a team that utilizes this Management Science tool in its operations.

Several chapters of this book appeared originally as a series of articles in *Industrial Canada*, a monthly publication of the Canadian Manufacturers' Association. A. W. House, editor of this magazine, incorporated numerous improvements into the series, and naturally these have carried over into the book.

For calculations pertaining to the illustrative examples and case-problem data, I am indebted to Miss Gail P. Anderson, graduate student assistant, for her careful and accurate work.

NORBERT LLOYD ENRICK

KENT, OHIO

CONTENTS

16 CASES CALLING FOR SIMPLEX ANALYSIS   153

17 CASES SOLVABLE BY THE STEPPINGSTONE METHOD   181

Appendix

MANAGEMENT PLANNING

PART I

Basic Methods and Applications

1

INTRODUCTION:
Planning, Programming, and Coordination of Operations

Mathematical Programming, as a simple-to-use tool for the planning and coordination of operations, was explored briefly in the Foreword. A large variety of applications will be shown in this book, taken from production, sales, administrative, and financial areas of management.

One of the key uses of MP is in the planning and coordination of production and sales operations. Accordingly, in introducing the subject of programming, we will examine this particular area of management, its requirements, and the nature of the type of coordination called for. This approach should serve particularly to highlight the interlocking nature of problems of planning, programming, and coordination, thereby laying a foundation for the material to be brought in subsequent chapters.

The value of MP in production-sales coordination is felt especially keenly in firms that must produce a large variety of products, models, styles, and sizes involving a diversity of processing stages, machinery, and equipment and where, moreover, a number of new products are usually under consideration in market research, preliminary design, or more advanced development stages.

Expanding product lines, increased diversification, and heightened emphasis on innovation all call for planning based on the insights derived through MP analysis.

Self-Test

Before embarking on a program of MP utilization, a firm may ask: Do we need this technique? The answer will depend on individual evaluations of desirability and practical feasibility for the company concerned, but it may help in arriving at a decision to take a brief self-test.

The test here provided consists of a four-part questionnaire covering:

— Sales problems.
— Production problems.
— Inventory problems.
— General cooperation problems.

Each part has nine questions requiring a brief answer of "often," "sometimes," or "rarely." Complete the questionnaire *before* you study and apply the scoring procedure.

It will be seen that the points scored are in the nature of "demerits" for the system now in use in your organization. But there is no reason to feel glum if you pile up an impressive total. This is the usual pre-MP situation, prior to the installation of MP and related new sales-production coordination methods. In fact, when the author circulated this questionnaire at a management meeting in Montreal he was more than once asked how it was that he knew so much "about the inside workings of our place" without having been there—a comment which serves to emphasize the universal and common nature of the problems noted.

Overall Score

An overall rating can be built up from the individual scoring points. A need for better planning of sales and production facilities may reasonably be considered to exist when:

1. For any *one* of the four parts, the score exceeds 6 points;
2. For all four parts *combined*, the total score is over 20 points.

If the test is taken by several management people, an arithmetical average of the scores may be used as a criterion of whether or not improved sales-production coordination methods should be adopted. (One executive's comment: "If too many raters in an organization come up with a low total of demerit points, they may not be very perceptive.")

The chief value is not, however, in the scores obtained. Rather, the self-test is valuable for the heightened consciousness it produces of the need for good sales-production coordination.

Thus, in a sense, the self-test lays the foundation on which the MP program can be developed. By taking the test, we have begun to install MP.

Table 1-1 Self-Test for Mathematical Programming Need

SALES PROBLEMS	Often	Some-times	Rarely	Score
1. Salesmen tend to promote unprofitable items.				
2. Unrealistic delivery dates are promised.				
3. Little effort is made to promote those items on which a sizeable inventory is on hand.				
4. Salesmen weak in maintaining prices.				
5. No regular checks made by sales management with production as to which items to promote.				
6. Laxity in evaluating costs and in pricing policies.				
7. New models late for season.				
8. Large quantities of distress merchandise at end of season.				
9. Orders accepted for uneconomical lot sizes and production runs.				

PRODUCTION PROBLEMS

	Often	Some-times	Rarely	Score
1. Too many changes in set-ups, models and styles				
2. Machine efficiencies low.				
3. Production bottlenecks and waiting lines.				
4. Excessive slack time				
5. "Current status" information hard to get when customer inquires about his order.				
6. Many "rush" orders.				
7. Customer cancels out because of late delivery.				
8. High costs on some lines.				
9. Below-standard quality on some lines.				

INVENTORY PROBLEMS

	Often	Some-times	Rarely	Score
1. Excessive raw stocks on some lines, not enough on others.				
2. Poor control of finished goods inventory.				
3. Large amounts of slow-movers in storage.				

	Often	Some-times	Rarely	Score
4. Shortage of fast-selling items.				
5. Spoilage, deterioration and other losses in inventory.				
6. Inadequate supplies.				
7. Lack of repair parts keeps machines idle or inefficient.				
8. Sales are lost because of "no inventory", and later the goods are discovered to have been in stock all along.				
9. Scattered responsibility for inventories.				

GENERAL CO-OPERATION

	Often	Some-times	Rarely	Score
1. Salesmen dislike "mill attitude;" use expressions such as "slow" and "unreliable."				
2. Some salesmen get more co-operation from plant than others.				
3. Production complains that sales is pushing unprofitable items that fail to adequately use plant capacity.				
4. Production protests that there are too many "rush" and "special" orders.				
5. Production is plagued by excessive diversification and too many short runs.				
6. Salesmen have inadequate appreciation of production problems.				
7. Production people have inadequate appreciation of sales problems and customer requirements.				
8. Not enough data on production, sales and inventories to make intelligent decisions.				
9. Data on production, sales and inventories are not up-to-date for quick decision-making.				

SCORING METHOD: If answer is "rarely", apply **0** points; if "sometimes", **1** point; if "often", **2** points.

The Need for MP

We have noted that more than ever in the history of manufacturing, industrial plants are being called upon to produce a large variety of products, models, styles, and sizes for both existing and new lines of merchandise. Problems of production-sales coordination arise because for *each* of these multitudes of products:

1. Production requires several operations, including various set-ups, speeds, raw materials, and processing states; and
2. Sales and related promotional efforts must be adapted to take account of market potentials, pricing factors, and the plant's capacity to deliver on the basis of various delivery time limits, quantity specifications, and cost considerations.

Meshing of these production and sales factors is therefore quite difficult, yet the success of overall business operations may well depend upon a full quantitative analysis that properly weighs and balances these factors. An example will illustrate this point. A particular product may show a good margin of profit per unit sold, and yet a more complete study may prove that it is uneconomical to make the product in large quantities, because its manufacture creates serious bottlenecks in processing. The bottlenecks may be in a department or on a critical machine, and may disproportionately limit production of other items, which, while perhaps less profitable per unit, contribute more to overall profit when total output is considered.

Nature of MP

Mathematical Programming (MP) serves to weigh and balance the multitudinous factors of profit margins and potential production bottlenecks in their effect on overall business operations—a feat so complex that the unaided human mind could never match it. The resulting information then helps sales and production management to find the optimum combination of products, models, and styles in such quantities as will maximize total profit.

We cannot, of course, attain this theoretical optimum in actual operations. Nevertheless, we can arrive at a mathematically determined program that would lead to an optimum and which therefore can serve as a *goal*, toward which selling, marketing, and productive efforts can be directed. Moreover, the program so obtained will give detailed—not just general—guidelines, which will be of daily practical use to all sales and production people.

MP is similar in nature to other quantitative decision-making tools: it develops information and guidelines that will be of service to any astute management group. Since these guidelines are in turn based on data and sources that come through management, management continues to be a crucial factor in the successful installation and implementation of all business-scientific programs within an organization. As quantitative methods and their uses increase, there will be increasing demand for really competent and effective managers at all levels of business and industrial operations. It is generally considered that such managers need not know the mathematics of many modern management aids, but they ought to be familiar with the general processes by which these methods develop information and guidelines, be able to assess their value and limitations, and be competent to install, implement, and use these techniques.

Summary

Mathematical Programming as a planning and coordinating tool finds some of its most useful applications when production and sales activities of an organization are to be meshed effectively. The general needs for such gearing together of two principal organizational functions and the nature in which MP helps management in this endeavor have been discussed. In the next chapter a simplified illustration will be given. A quick glance at subsequent materials will serve to demonstrate that MP has many further valuable and often crucially important applications. Production-Sales coordination is merely the first area of application considered in this book.

2

HOW PROGRAMMING WORKS

In order to understand how MP works, we may follow a greatly simplified example, shown in Table 2-1 and supplemented by Figs. 2-1 to 2-4. Our hypothetical plant produces two products, A and B, with different contributions to profit (also known as "operating profit" or "variable margin"); two processing stages, Machining and Assembly; and different production rates for each product and department.

Data on profit-contribution per unit come from sales expectation of prices and cost estimates, while production rates are derived from time study and incentives data. We note that product A brings the higher profit ($10 per unit) compared to product B ($8 per unit). Although there is enough productive capacity to machine 40 units of A, only 20 units can be assembled. Assembly is thus a bottleneck that limits output of A to 20 units. On the other hand, a bottleneck in Machining limits output of B, so that only 30 units can be produced.

The production rates just quoted appear in line c of Table 2-1, and they are based on the assumption that *only* product A *or* product B is made. Correspondingly, expected profits are:

1. For A, 20 × $10 or $200 per week; and
2. For B, 30 × $8 or $240 per week.

Therefore, if only one product is to be manufactured, B would be preferable.

Since both products have different bottleneck processes, the next step is to examine a combination of the two products, balancing the two bottleneck processes and seeing whether we might not reach a higher profit than the $240 of line d.

Finding the Optimum

In order to find the combination that will maximize profit, a graphic analysis will be employed. As a first step, a graph as

Table 2-1 Solving a Simple Problem of Product Programming

Problem data	Product A	Product B	Productive capacity hr/wk	Total
a. Profit contribution, dollars per unit	10	8		
b. Production rate, hours per unit				
Machining	6	8	240	
Assembly	4	2	80	
c. Output if *only* Product A *or* B (not both) is produced, units per week				
Machining*	(40)	30†		
Assembly	20†	(40)		
d. Profit, dollars per week Found by multiplying $a \times c$	200	240		240‡
Problem solution				
e. Optimum product combination, units per week Found from Figs. 2-1 to 2-4	8	24		
f. Profit, dollars per week Found by multiplying $a \times e$	80	192		272§

* Productive capacity divided by production rate. Thus weekly machining capacity of 240 units, divided by production rate of 6 hours per unit of A, gives a machining output of 40 units per week.

† Bottleneck in processing for this product. Thus, even though theoretically we can machine 40 units of A, only 20 units can be assembled. Assembly is therefore a bottleneck for Product A. For Product B, Machining is the bottleneck.

‡ Only A or B, not both, can be produced at this stage, so that maximum profit corresponds to Product B at $240 per week.

§ Optimum for combination of both products, obtained from addition of $80 and $192 for A and B, giving $272. This amount is better by $32 or 13 percent than the previous $240. No other product combination will give a higher profit under the profit and production factors given by the Problem Data.

shown in Fig. 2-1 is constructed with quantities of product A on one base line and B on the other. Now connect the 40 on A (representing the units of weekly capacity in Machining if only

A is produced) with the 30 on *B* (representing the units pro-
duceable in Machining if only *B* is made). Next, connect the 20
on *A* with the 40 on *B*.

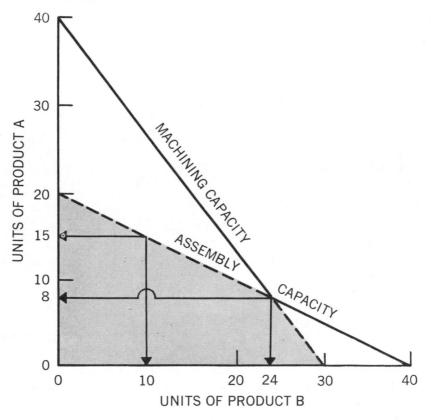

FIG. 2-1 First step in finding optimum product mix. Machining capacity line con-
nects 40 units of Product A with 30 units of Product B, corresponding to weekly
capacity (Table 2-1, line c). Assembly capacity is also drawn in. Shaded area represents feasible production. Beyond this area, bottlenecks in machining or assembly
prevent further output. Some of the feasible outputs are: 20 units of A only, 30 units
of B only, 15 units of A and 10 of B, 8 of A and 24 of B, as indicated on the graph.
Problem is to find the optimum for overall profitability.

We have now shown the two lines indicating the capacity
limitations set by Machining and Assembly. But we also know
that bottlenecks make it impractical to machine more than 20
units of *A* or assemble more than 30 units of *B*. Only the shaded
part of the graph, therefore, represents feasible product combinations. For example, if 10 units of *B* and 15 units of *A* are
made, available assembly capacity is exhausted. (See arrows

leading to B of 10 and A of 15 units on graph, which are bordered by the "assembly capacity" line.) Similarly, 24 of B and 8 of A will exhaust both assembly and machining capacity.

The information shown may also be given in cubic form, as illustrated in Fig. 2-2. This method has the advantage of permitting addition of a further variable—the profit. The profit structure (Fig. 2-3) will reveal at a glance the total profit per week associated with various product combinations. For the forementioned 10 units of B and 15 units of A, the total profit is $10 \times \$8 + 15 \times \$10$ or $\$230$. This amount is higher than the $\$200$ for A alone, but less than the $\$240$ for B alone. The graph also emphasizes, as expected, that production below capacity is undesirable. For instance, if we produce 8 of A and 10 of B, we earn a meager $\$160$ per week of profit. We are inside the shaded area, away from its borderlines.

Looking at the trend of heights of the dollar columns, a promising point is noted at the intersection of Machining and Assembly capacity, corresponding to 8 units of A and 24 of B. This point yields a $\$272$ profit ($\$10 \times 8 + \$8 \times 24$). Any movement away from this product "mix" or combination yields less in total profit returns (see Fig. 2-4).

Investigating Corners

This simplified problem has demonstrated a crucial principle, valid for all mathematical programming involving linear relationships among the variables: in order to find the optimum, investigate the corners.

By analyzing the points 0, 20 on A, 30 on B, as well as the intersection of the two capacity lines, we have constructed a dollar profit dimension which also reveals the optimum point.

Maximum profit, however, will not always involve a combination of products. For example, if product B had had a profit of $\$15$ per unit, then production of 30 units of B would only yield $\$450$ total profit, which is more than a combination of 8 units at $\$10$ for A ($\$80$) plus 24 units of B ($25 \times \$15 = \$360$), since $\$80$ plus $\$360$ comes only to $\$440$.

The principle of investigating corners is also applicable to problems involving a large number of products and processing stages. But there is a problem: with each new product included in the analysis, one must add a further dimension to the space-structure to be investigated. For two products we needed three dimensions (a cube), for three products we will need four dimensions, and for 100 products 101 dimensions are required. Since human beings can conceive only three-dimensional space,

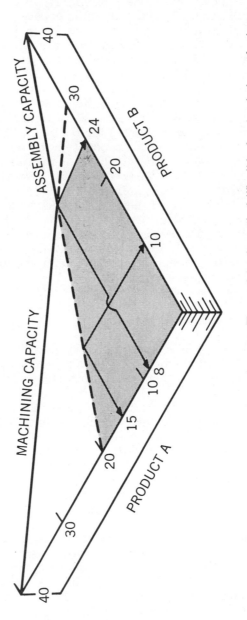

FIG. 2-2   Second step in finding the optimum product mix. The graph has been tilted "flat," by forming the base of a three—dimensional (cubic) visualization. The third dimension is needed to add the profits associated with each product output.

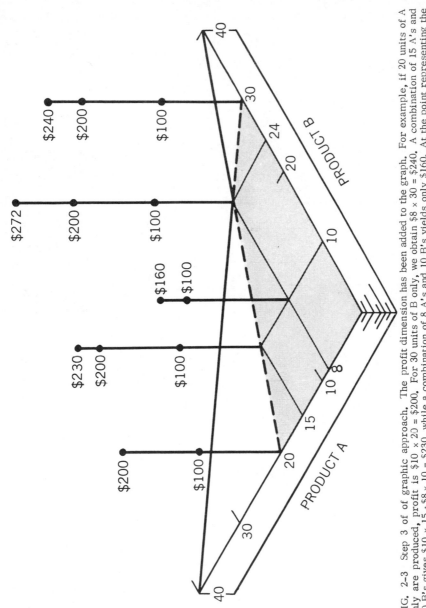

FIG. 2-3   Step 3 of of graphic approach.   The profit dimension has been added to the graph.   For example, if 20 units of A only are produced, profit is $10 × 20 = $200.   For 30 units of B only, we obtain $8 × 30 = $240.   A combination of 15 A's and 10 B's gives $10 × 15 + $8 × 10 = $230, while a combination of 8 A's and 10 B's yields only $160.   At the point representing the intersection of the capacity limitation lines, corresponding to 8 A's and 24 B's, profit is $272.   From a study of this graph, we conclude that the optimum product output will be found from one of the corners of the feasible production base (shaded area). The three corners in question are 20 A only, 30 B only and 8 A plus 24 B.

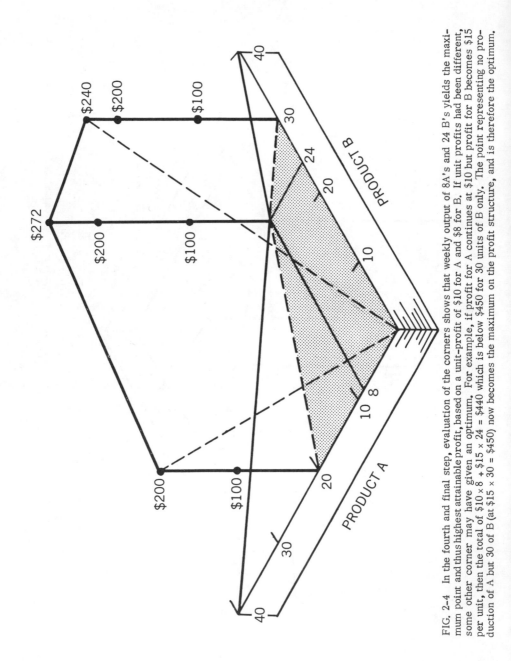

FIG. 2-4 In the fourth and final step, evaluation of the corners shows that weekly output of 8A's and 24 B's yields the maximum point and thus highest attainable profit, based on a unit-profit of $10 for A and $8 for B. If unit profits had been different, some other corner may have given an optimum. For example, if profit for A continues at $10 but profit for B becomes $15 per unit, then the total of $10 × 8 + $15 × 24 = $440 which is below $450 for 30 units of B only. The point representing no production of A but 30 of B (at $15 × 30 = $450) now becomes the maximum on the profit structure, and is therefore the optimum.

we must resort to mathematical models in investigating corners in large-scale analyses.

We referred to the fact that large-scale models require mathematical analysis. The particular procedure utilized for sales-production coordination problems of the type discussed here and in the next chapter is the so-called Simplex* method of matrix steps. It is presented later, in Chap. 14, after the practical operational uses of MP have been given. An understanding of this mathematical background is not needed for practical applications by the manager or executive. Once the MP problem has been formulated properly, these mathematical steps toward a solution will, in any case, be performed on a computer.

Summary

Utilizing an illustrative example as a vehicle, we have examined the principles of Mathematical Programming, as applicable to the solution of a problem involving interlocking restrictions that affect the attainment of a goal. We are now ready to examine a case-history of a full-scale application.

---

*The Simplex method was originated by G. B. Dantzig, and first published as Chapter 12 of T. C. Koopman's (editor) *Activity Analysis of Production and Allocation* (New York 10016: J. Wiley & Sons, 1951), Cowles Commission Monograph No. 14.

3

FULL–SCALE PROGRAMMING:
A Case History

Finding the optimum combination of products, models, and styles which add up to maximum overall profitability is not always an easy task under today's highly competitive market conditions. For this reason, the recently developed methodology of Mathematical Programming (MP) is being used increasingly as a quantitative-analytical aid to management in coordinating sales and production tasks.

The necessity of MP analysis for astute management was brought out previously. The illustration involved only two products (*A* and *B*) and two processing stages, and thus a three-dimensional (cubic) diagram easily showed the optimum point. For more than two products, however, we need a multidimensional analysis which can be carried out mathematically but which can no longer be graphed (since humans cannot visualize more than three dimensions).

Handling the Multiproduct Problem

A medium-sized MP application may involve some 50 to 150 products and 15 to 25 processing stages. Using a desk calculator, it is possible to work out a program of desirable products and quantities, but this will take about two weeks of concentrated effort. Fortunately computer programs are available. The transfer of the basic data—products or models under consideration, prices and profit-contributions expected, production rates, productive capacity, and other data of significance—to punched cards or tape is a matter of relative simplicity. This data, once fed into the programmed computer, will produce the optimum solution within a matter of minutes.

Two factors should be kept in mind:

1. You need not have an electronic computer "in place" in your organization. Various time-sharing or service arrangements are available today, whereby a computer solution is obtainable for anyone at a quite low cost.
2. Many computer programs for solving MP problems exist, and it is a simple matter to select the one that is needed for a particular application.

Illustrative Example

The case history that follows is taken from experience with a medium-sized MP application. Space limitations, however, make it difficult to present all of the 153 products (including models and styles) and 27 process departments. The lesson learned from this application, however, can be brought home effectively by showing only four of the principal products and four of the key processing departments (in Table 3-1).

In the Problem-Data section, we note the unit-profits per model (from C to F), the production rates at which the metalworking, wiring, assembling, and testing departments can process each unit, and the productive capacity in each department. Next, in line c we observe the Optimum-Profit solution in terms of quantitites to make; followed by the resultant total profit per model together with the profit total. Lines c and d thus contain the problem solution. The usage of the available capacity, made by this optimum quantity selection, is analyzed in c. All departments, excepting wiring, have been employed to capacity.

Sales Manager's Objections

When MP was first proposed in this organization, the sales manager voiced doubts about its value. Looking at the type of data shown in line a of Table 3-1, he raised the objection: "It is easy to see that we make the highest profit on product C and the lowest on product F. No mathematics is needed for that. The problem is that people don't buy much of product C, and we are already spending too much money and effort promoting this hard-to-sell item."

Valid as these objections may be from a strictly sales viewpoint, it was possible to show, by means of the MP analysis, that sale management could profit greatly from this technique.

Table 3-1 Mathematical Programming Problem and Its Solution

Problem data	Product C	Product D	Product E	Product F	Productive capacity hr/wk	Total
a. Profit, $/unit	.80	.74	.72	.56		
b. Production rate in Hours/unit						
Metalworking	.306	.306	.326	.224	250	
Wiring	.720	.648	.720	.792	720	
Assembly	.360	.288	.288	.216	242	
Testing	.180	.260	.240	.290	260	
Problem Solution						
c. Optimum product quantities, units/wk	56	379	0	522		
d. Profit, $/week = $a \times c$	45	280	0	292		617*
e. Production hr/wk used by optimum product quantities = $b \times c$						
Metalworking	17	116	0	117		250
Wiring	40	246	0	413		699**
Assembly	20	109	0	113		242
Testing	10	99	0	151		260

*Total profit is found by adding the profits-per-week for each product, in this line. No other quantities of output for the four products under consideration can yield a higher profit.

**Comparison with corresponding productive capacity (line *b*) shows that wiring (with 699 hours of use) is the only process not using all of the productive capacity (of 720 hours).

Note also that Product *F* with lowest profit-per-unit is nevertheless the most desirable product, by contributing most to total profit (compare lines *a* and *d*), while Product *C* with highest unit-profit is relatively much less desirable (by creating production bottlenecks).

A clue to this fact can be gained from a review of the plant manager's attitude.

Plant Manager's Position

The plant manager felt that some weight should be given to production balance. For example, while product *C* did yield a

high unit profit, it was also likely to create a bottleneck in Assembly. With bottlenecks occurring, some machines and departments cannot meet demand with overtime runs, while other
equipment stands idle at least part of the time. Therefore, unbalance of productive facilities (machinery, equipment, skilled
operators) is an important cost factor that may often offset any
gains from high unit profit.

Solution to Sales-Production Problems

In seeking a solution to the two types of problems—sales and
production—just noted, an MP program was run on a computer.
This computer analysis posed the question: "Which products in
what quantities will result in an overall maximum profit for the
company?"

Using a medium-sized computer, a solution was obtained in
a matter of minutes, showing the type of information in line c of
Table 3-1. (Our discussion is in terms of four key products,
even though actually many more items were involved.) In particular, line c shows optimum product quantities that match
sales potential with productive capacity. From line d, the weekly
profit attainable if this optimum can be attained, is found to be
$617. No other combination of product quantities will result in
a higher profit.

A Surprising Result

As happens in most MP work, the information revealed was
quite surprising to the managers. Product C, which the sales
manager had considered a desirable (though hard-to-sell) item,
was actually a poor contributor to overall product. Even if it
were not hard to sell, it would be of relatively little value to the
overall production program. (Products D and F are better, as
line d shows.)

Not only did MP "knock out" the highest unit-profit item as
a desirable product to push, it also showed that product F (at
only 56¢ unit profit, as against C's 80¢) was a good item, and
actually represents the greatest contributor to overall profit.
While the sales department had neglected this product, it now
realized that F was actually a quite profitable item. Sales
effort was shifted from C to F. By similar shifts of emphasis
for all of the company's products, a net gain of twenty percent
in overall profitability was achieved in the next 13-week quarter.
Moreover, MP became a routine part of management's overall
information system, for use by both the sales and production
executives concerned.

Gaining a Perspective

Management gained a completely new perspective of its marketable products. In the actual analysis involving all the products, many past favorites lost out against previously over-looked items. As the routine of MP became fully understood, sales and promotional efforts could be directed far more intelligently and decisively.

In addition to higher profits from increased sales, the plant also gained in morale as a result of better balance of production and smoother flow of work, with less overtime and part-time operations. In this connection it is of interest in reviewing the scheduling information (line e of Table 3-1) that metalworking (blanking, pressing, and machining), assembly and testing use up available capacity, while in wiring operations there is a slight but unavoidable slack.

Today many companies perform weekly MP analyses, since not only the day-to-day problems of sales and production, but also long-run planning phases of management can benefit. Examples occur in these areas:

— Considering the installation of new equipment in processing departments that may be limiting overall plant capacity.
— Emphasizing the need for automation in operations where a chronic shortage of skilled operators is experienced.
— Preparing for long-term expectation in technological and market developments.
— Integrating other tools of quantitative management information and control, such as inventory control and market research, with the MP program.

The word "programming" is conventional but unfortunate, in that it implies a rigid or fixed plan. Yet, MP is merely a means of analyzing a large number of interrelated factors of marketing and production, with the aim of gaining an overall perspective. From this vantage point, in which the individual factors have been weighed in terms of their contribution to total profit, better executive decisions can be made. These decisions, in turn, will affect all production, marketing, and related activities of the company.

The resulting plans and programs are usually quite *flexible*. They serve as *guides only*, subject to review and revisions, based upon actual market and product developments. And, of course, at all times the MP findings are translated into action in terms solely of the judgment of the executives involved in sales, production, and related activities. Used as a flexible guide, MP is thus not a rigid schedule but rather a source of information in managerial decision-making and control.*

Summary

This case history has demonstrated the value of Mathematical Programming in management planning. Next we should consider how MP is best applied within an organization, as will be done in the next chapter.

*This aspect of MP is further illustrated in a report on the introduction and use of sales-production coordination at Pepperell Manufacturing Company. See "Simplex Programming: A Case History" by K. Eiseman and W. M. Young, Chapter 8, contributed to N. L. Enrick's *Management Operations Research* (New York 10017: Holt, Rinehart & Winston, Inc., 1965).

4

APPLYING PROGRAMMING
IN AN ORGANIZATION

For the purpose of examining the application of MP in an organization, we will use as the frame of reference the type of planning for production-sales coordination previously presented in connection with the illustrative case-history of a multiproduct manufacturer involving several processing stages.

Initial Steps

Before MP can be applied on an organization-wide or even plant-wide basis, the following initial steps should have been taken:

1. The nature of MP and its potential usefulness have been presented to all important persons in management concerned, particularly to those who are directly involved in goals, plans, programs, and management of productive and sales efforts.
2. The results of trial evaluations of MP have been submitted, utilizing the specific cost, production, and sales data pertinent to the organization.
3. While not everyone has been fully sold on MP, the majority of the executive, managerial, and supervisory people concerned understand its nature and are willing to try and work with it.
4. For a period of several weeks "dry-run" data sheets on the results of MP studies have been submitted to management, and some use of this information has been made in directing sales efforts and in planning production schedules.

Usually these four steps will involve six to twelve weeks of preparatory work, and the sequence of steps as well as their individual detail will vary considerably, depending on the particular requirements of individual instances. During this period it will be useful to have information—such as these articles— available as general background material. MP, admittedly, may look revolutionary *at first* and the degree of effort involved in laying a proper groundwork and even in educating those concerned about the aims of the program should not be underestimated. However, once this work has been accomplished, we are ready to apply MP in earnest for the benefit of the company.

Information Flow

Mathematical programming, in essence, is a system for providing management with information as aids in decision making. Effectiveness of the system hinges to a large extent on how adequately and smoothly the important information flows. The following is an example of this flow which was found very successful as applied in actual practice:

1. **Determining Products to Stress in Sales Program.** Based on data regarding 1) productive capacity, 2) market developments, 3) expected prices, and 4) expected costs and thus expected profits, the computer prepares an initial MP analysis.

2. **Information in Initial Analysis.** From the initial analysis, sales management will note the goal quantities for each product style which, if attainable, would optimize profits. Production, utilizing this same information, notes what production scheduling will be needed.

3. **Revision.** Based on step 2, it is not unusual for sales and production to find that some of the initial information on productive capacity and related production factors, as well as on saleability, marketability, and pricing possibilities, was incomplete. The revisions are noted and a new MP analysis is prepared accordingly. As experience in the use of MP is gained, the need for such revisions in the basic data will diminish.

4. **Routine Preparation.** Eventually, MP analyses—prepared by the planning or other department assigned—become routine. Usually, it is best if the MP analysis work is done on a Saturday, with the pertinent results becoming available to sales, production, and other management concerned by early Monday morning.

5. Routine Use. Sales utilizes the MP results in planning sales activities, promotional work and marketing strategies. Production works closely with sales in scheduling equipment, providing for overtime where required, or scheduling purchase of supplies, components, and parts in accordance with expected sales. In instances where long-term needs are foreseen, the purchase of additional equipment may require consideration.

6. Sales and Production Adjustments. During the week, as sales occur and production progresses, provision must be made for continuous adjustments of pertinent data on 1) unsold stock on hand and 2) uncommitted production facilities.

7. Price Changes. Decisions regarding price changes must be noted. These decisions, together with any changes in cost factors, will affect unit profits. Computer data on prices, costs, and profits will need revision accordingly.

8. Other Adjustments. Record and enter other pertinent developments that must be allowed for in the computer program. Program revisions may become necessary as new and previously unforeseen factors become of significance for sales or production planning.

9. New Data. During the weekend the computer uses the information from steps 6, 7, and 8, to prepare new up-to-date data on 1) stock on hand, 2) sales records, and 3) production expectations for sales and production purposes. The determination will result in a new MP analysis of the most desirable products and quantities to sell or to stress in the promotional program.

10. New Cycle. With preparation of the new data and analyses from step 9, the next week's cycle of sales promotion, production scheduling, and related business operations begins. Moreover, while these weekly activities occur, the MP analyses will also have afforded a view of long-term trends and requisite longer-term planning.

Executives and Managers Make Choice

In considering the flow of MP information, it should again be emphasized that the system still leaves to the manager-executive the problem of decision-making. However, the critical nature of decisions has been sharpened. Why? Because MP emphasizes

the fact that in judging the feasibility of any proposed program—
sales goals, production facilities, pricing, etc.—the responsible
decision-maker must consider the alternate methods of using
the company's resources.

For example, the sales manager in deciding on diversification of product lines becomes more keenly aware than ever that
his decision may be quite two-sided. Diversification may increase the sales potential, in that a wider market is now available. On the other hand, this diversification may be purchased
at prohibitive costs in terms of tie-up of productive capacity to
accommodate an excessive number of changes. In turn, once
productive factors have been considered fully, the cost of a
highly diversified product line may lead to product prices that
are too high in relation to the market.

As another illustration, the production manager may be
overly cost-conscious and rebel against any attempt to diversify.
Here, MP may demonstrate that a certain amount of diversification is essential if an adequate sales volume is to be maintained.
In this conflict of "streamlined production operations" vs
"diversified product line," an MP analysis showing the road
toward the optimal degree of diversification may be extremely
valuable to the decision-makers.

Convincing the Skeptic

Winning over the skeptic is part of the job of a successful
MP installation. Let us examine some frequent objections to
MP and the answers to them:

Objection: "In MP analysis, lightning-fast computers run by
crew-cut young 'wizards' are making major business decisions."

Answer: Though computers are important for MP analysis,
the analysis itself does not play a paramount role in the decision process. It merely helps the decision-maker to sharpen
his judgment in choosing the best among several alternative
possibilities of action.

Objection: "MP analysis is a substitute for good, sound
management judgment."

Answer: True enough, MP does provide a mechanism for
cranking in production and sales data for an overall assessment
of all important interacting factors. All these are quantitative
in nature. Other factors, however, such as "How will the company

image be affected?'' or ''What will be the effect on operator morale?'' are questions involving qualitative judgment. Yet, the need for consideration of qualitative aspects of a decision does not mean that quantitative factors ought to be discarded. The more informed we are on quantitative factors, the better we are able to gauge the qualitative effect of decisions. MP analysis will thus upgrade, not downgrade, the judgment element in decision-making.

Objection: ''MP analysis is prejudiced against established patterns and favors the introduction of new approaches.''

Answer: If this is fact, then let us recognize that such prejudice is not necessarily an unfair one. Actually, of course, it is ''improved organizational effectiveness'' and not any battle of ''new'' vs ''old'' that merits attention.

Objection: ''MP as a system is only as good as the information fed into it.''

Answer: This is true. MP analyses have occasionally failed because not enough time was invested in working with all persons in the organization who could provide pertinent and valuable information on the types of input-data that were needed. On the other hand, an alert management will quickly recognize flaws in input-data sources, so that corrective action will be taken before MP is put on a full-scale routine basis.

Summary

We have observed that routine MP, in practice, is not a ''decision-making process.'' It does not say ''Yes'' or ''Go-ahead'' to any plans, schedules, programs, or projects in the production-sales area. On the other hand, MP analysis has come into its own as a valuable adjunct to informed, responsible decision-making in the modern management process. This position will unquestionably become more firmly established as management learns to make better use of MP.

5

DYNAMIC DECISION—MAKING CRITERIA

When an organization decides to install mathematical programming as a management aid, it will often happen that the standard procedures can be readily adapted to the individual company's needs. In other instances, however, modifications in varying degree may be required in order to fit special factors that affect either marketing or production or both.

This chapter will illustrate a situation in which a major modification of MP was required. This example is interesting because it resulted in the development of highly flexible decision-making criteria under dynamic marketing conditions. Beyond this, it serves to emphasize the fact that in general it should always be possible to develop suitable modifications of standard programs to fit the specific needs of effective management.

Development of Dynamic Decision Criteria

The manner in which dynamic decision criteria were developed, in our example, is not untypical of the general process by which an MP application becomes introduced in a company. The following occurred:

1. After an initial presentation of the merits of MP in sales-production coordination, management authorized a feasibility study.
2. For the purpose of this study, data were gathered for the various models produced by the company, showing:

   a. Direct and indirect costs.
   b. Anticipated marketable volumes and prices.
   c. Variable-margin or contribution to profit and overhead.

    d. Standard production rates, in hours per unit.

    e. Available productive capacity in terms of total machine hours, less machine hours committed on long-term contracts and thus not available for further immediate planning.

Production, Sales, Accounting, and Standards were the principal departments which cooperated in obtaining these data.

3. An MP analysis of the interrelated aspects and significance of the production, cost, and profit was then prepared. Although a considerable number of product models was involved, it will be sufficient to show an abbreviated example of only three models, as given in Table 5-1, lines *a-c*, supplemented by the further information in lines *e-f* (note that line *d* was not supplied at that time).

4. As explained to management, the data in Table 5-1 showed that for the three models weekly production of 7.4, 10.4, and 12.3 units, respectively, would result in maximum profit of $1905. This optimum program thus serves as a goal for guiding the efforts of sales promotion and production planning activities.

Although these results were found interesting and valuable by management, it was felt that the MP solution would not be fully applicable in a dynamic marketing situation, such as the company was facing at each of its two selling seasons that reached their peak in the spring and in the fall. Here is how the sales manager described the problem:

A large customer may be at our sales office willing to place a sizeable order. Often these offers may occur early in the season, so that we will have available an uncommitted capacity. It is then necessary to make a long distance call to the plant to present the offer to the general manager. Now, what should be done if the offer involves a quantity in excess of the optimum? For example, if the customer's offer involves 13 units of the "Z" model per week, this exceeds the theoretical optimum of 12.3 units by 0.7 units. If we accept this offer, what will it do to us in the way of unbalancing production facilities in relation to the optimum program (line *c* of Table 5-1)?

Further discussion brought out the fact that there were many aspects in the decision-making process, but a key item was the extent to which any manufacturing commitments might unbalance plant production away from the optimum goal. A simple answer might be that the effect of this unbalance could be evaluated by performing a new MP analysis based on the changed quantity for the "Z" model. In practice, however, there is no time to wait for such a new analysis. The customer is in the sales office, ready to sign an order, and the discussion with plant is in process over the long-distance wire.

A better answer to this problem was to develop the concept of dynamic decision criteria, DDC.

Nature of Dynamic Decision-Making Criteria

The dynamic decision criteria are indicated in line d of Table 5-1 and represent a set of 95% criteria, as chosen by top management. For each model, the DDC answers this question:

If total profits are to remain at least at 95% of the ideal attainable, then how much more than the optimum of 12.3 can we accept for weekly production of the "Z" style?

We recognize that production per week at a higher level than that recommended for any one model will encroach on available capacity for other models, thus reducing the overall optimum ideally attainable for all models. The ideal optimum is $1905 for the data under discussion (in Table 5-1). A "95% of optimum" level would represent a 5% loss and a total profit per week of only $1810.

As line d indicates, we can deviate from the goal program by producing 1.7 units per week of the "Z" model in excess of the optimum of line c, and still remain within 95% of the overall ideal optimum. Since the customer's offer is only 0.7 units above the 12.3 units recommended in line c, and within the allowable 1.7 additional units provided by the DDC, we can accept it.

Practical Result

The DDC provides a guide for judgment, in addition to the guides normally provided by an ordinary MP analysis. As with all guides, they merely serve management in the decision-making process. For example, despite the MP data and the DDC allowances, there may be other reasons—such as latest information regarding market and price developments, a special desire to please certain good customers, and similar considerations—which may cause management to decide against its own general guideposts in specific instances requiring special action.

Table 5-1 Problem-Example with Optimum Profit Solution
and Dynamic Decision Criteria

Problem data	Model "X"	Model "Y"	Model "Z"	Productive capacity hr/wk	Total
a. Profit, in $/unit	90	60	50		
b. Production rate, hours per week					
Metalworking	6	18	30	600	
Assembly	8	28	4	400	
Finishing	12	6	4	200	
Problem solution					
c. Optimum quantity for each model, units per week	7.4	10.4	12.3		
d. Dynamic decision criteria, units per week*	1.5	0.4	1.7		
e. Profit ideally attainable = $a \times c$, in $	666	624	615		1905**
f. Production hours used per week ($b \times c$)					
Metalworking	44	187	369		600
Assembly	59	291	49		399***
Finishing	89	62	49		200

* These criteria have been set at 95% of ideal, thus resulting in an expected profit of 0.95 × $1905 = $1810.

** Total is found by adding the profits for each model. No other quantities of output for the three models under consideration can yield a higher profit.

*** Because of rounding, this total of 399 does not fully equal the 400 hr in *b* above. Although for all processes production hours used equal capacity available, this is not necessarily a requirement for optimum profit.

This, however, does not detract from the value of MP and the DDC. Moreover, whatever decision is reached can be arrived at more quickly with MP and DDC information on hand as a ready reference.

Management will, of course, determine the level at which DDC's are set. In the present example it was indicated that a 95% level was used, representing management's willingness to make allowances of up to 5% in relation to the ideal optimum. Such an allowance level is usually expected in times of good market demand, with the expectation that, as the season gets under way, there will be little difficulty in selling all that available capacity can produce. On the other hand, when times are relatively slow as regards market demand and future expectations, management may decide that a level as low as 80% will be appropriate for the DDC calculations. It is apparent that, under weak market conditions, an early order turned down (because it happens to fit poorly into the program balance) could result in large amounts of unused capacity as the season progresses (if no substitute orders develop).

Thus, management must weigh the losses from nonoptimum acceptance of early offers against the cost of potential idle capacity as the season gets under way. Obviously the expected demand situation and related anticipations will thus play a major role in management's decision as to the level at which DDC's are to be established. In some instances, management may desire not just one level (such as 95%) but several levels (such as, say, 85, 90, and 95%) thus calling for several lines in the place of what is now in line d, of Table 5-1.

Some Technical Detail

DDCs are useful for on-the-spot decisions affecting acceptance or rejection of above-optimum quantities for a particular product, model, or style. Once a decision has been made to accept a customer bid, a new MP analysis is required to be performed as soon as possible. This involves the following detail:

1. A new MP problem is set up, by subtracting the contracted quantities (such as 13 units per week for the "Z" model) from productive capacity (column near right-hand margin of Table 5-1).
2. A new MP analysis is obtained.
3. The new MP analysis will contain a new row "c" of optimum quantities for an overall goal (ideal program), plus new DDC allowances.*
4. The new analysis and the new DDCs will serve management in the next instance when an on-the-spot decision must be made.

Requests for a large quantity of one model by a single customer are not likely to occur often. But when such offers do

---

*The matrix algebra involved in calculating DDCs was previously published in a relatively mathematical paper in *Management Science* (vol. 11, no. 8, June 1965) by N. L. Enrick.

come in, it would seem essential that quick decisions be arrived at, since failure to act promptly might result in loss of the order. The DDCs fill the gap. They are also likely to make MP more acceptable to management, by making the quantity recommendations of the program more flexible, thus avoiding the notion of a "fixed limit."

Summary

The DDCs were developed in response to a specific situation, but it may well be that *all* MP applications designed to coordinate sales and production will benefit by the inclusion of this concept, thus enhancing the value of this management technique.

6

RATIO—ANALYSIS METHOD

Once the merits of a full—scale MP application have been accepted, a question often asked is: "Do you have a desk-calculator method, one that is possibly less exact than the computer approach,* but that will do about the same job?" In answer to this, we will refer to a rather approximate method known as ratio-analysis. It should be pointed out, however, that despite its relatively few and readily applicable steps, the desk-calculator approach is probably more costly in terms of clerical time than would be charged by a computer for the precise solution.

Illustration of Ratio-Analysis Method

We will examine an illustration of the ratio-analysis technique in Table 6-1 and then compare the results with the findings from the more precise MP solution, in Table 6-2. First we will examine in step-by-step sequence the data pertaining to the ratio-analysis approach:

Line a: The profits in dollars per unit are shown for the saleable products, A, B, and C.

Line b: We now take the product with the highest profit, A of $9 per unit, and call it 1.00. Next we express all other profits in ratio form. In particular, for B, we have $9 to $6 as the ratio of A to B, giving 1.50 as the result. For C, $9 to $5 gives 1.8.

Line c: The usual production rates are shown for the three processes, together with the capacities. In column 5, each

---

*The "computer approach" referred to is known as the Simplex method of matrix steps. The term "Simplex" refers to a particular mathematical notion; it does not mean to imply that the method is simple to work manually.

Table 6-1 Programming Problem Evaluated by Ratio Analysis Method

Problem data and analysis steps	(1) Product A	(2) Product B	(3) Product C	(4) Productive capacity, mach.-hr. per week	(5) Ratio of maximum capacity process, (machining = 600hr) to other processes
a. Profit, $/unit	9.00	6.00	5.00		
b. Ratio of maximum profit to others	1.00	1.50	1.80		
c. Production rate, hr/unit					
Casting	12	6	4	200	3.0
Machining	6	18	30	600	1.0
Finishing	8	28	4	400	1.5
d. Percentage of capacity required for one unit					
Casting = c_1/c_4	6*	3	2		
Machining = c_2/c_4	1	3	5†		
Finishing = c_3/c_4	2	7*	1		
e. Percentage of capacity needed for equal profit, = $b \times d$					
Casting	6.0†	4.5	3.6		
Machining	1.0	4.5	9.0*		
Finishing	2.0	10.5*	1.8		
f. Percentage in e adjusted for relative capacity, $c_5 \times e$					
Casting	18.0*	13.5	10.8†		
Machining	1.0	4.5	9.0		
Finishing	3.0	15.8*	2.7		

*This is the bottleneck process for the particular product (A, B, or C) by being the maximum value (in casting, machining or finishing of A, B, or C).

†This is the minimax, that is, the minimum of the maximum values.

Result: In section f, the minimax is in casting, for product C. This is the most desirable product to sell in quantity, therefore. In that same row, products B and A show ascending bottlenecks of 13.5 and 18.0; thus next to C, product B and then product A are desirable.

process is shown in ratio form with regard to the maximum-capacity of 600 hr in machining. Thus, comparing machining to casting, 600 to 200 hr, gives a ratio of 3.0.

Line d: What is the percentage of capacity required to make one unit? Let us look at product A. It takes 12 hr to make one unit in casting, but there are 200 capacity hours available. This means that to cast one unit of product A requires $100 \times 12/200$ or 6% of capacity. All other percentages are obtained similarly. In making product A, note that the casting consumes the largest percentage of capacity (6 vs 1 in machining and 2 in finishing). Casting is thus the bottleneck for product A (assuming for a moment that only A is produced). For B the bottleneck is finishing, at 7%, and for C it is machining, at 5% of capacity respectively. Looking at the three maximums, 6, 7, and 5, we note that the last percentage of 5 is the lowest—the minimum of the maximums—which is also called the *minimax*.

Line e: We now ask the question: "How much of capacity is needed to make equal profits?" Line d gave us the simple percentages of capacity. But each product gives profit at a different rate, as line a shows. It is clear, therefore, that we must adjust the percentages in d to reflect these differences. Multiplication of the values in d by the profit ratio in b gives the desired answers.

Line f: A further adjustment is now made, by multiplying the ratios in e by the relative capacities reflected in column 5. We are now ready to look at the minimax. It occurs in casting, and product C impinges least on this bottleneck factor. C is thus likely to be a desirable product to make in relatively large quantity. In the same line, product B with 13.5 is next best as regards relative quantities, while product A with a ratio of 18.0 should be considered least desirable.

We have found an approximate solution to our MP problem. In the process, further insights were gained regarding the relation of unit profits, production rates, and productive capacities in their combined effect on ultimate quantities that will yield optimum profit. But the solution cannot tell us just how much to produce of each item.

Precise MP Method

It will now be desirable to compare the Ratio Analysis findings with the results obtainable by more precise procedures, using the full-fledged MP technique as shown in Table 6-2. We find:

1. Both methods recommend the same relative quantities. Note that in line c, Table 6-2 again shows C, B, and A in this order of desirability: viz., recommending that 12.3 units be made of C as against only 7.4 units of A per week.
2. Table 6-2 gives exact quantities, while Table 6-1 gave only relative desirabilities.
3. Table 6-1 gives certain insights of the structure of relationships not revealed by Table 6-2.

Table 6-2 Programming Problem and Its Exact Solution

Problem data	Product A	Product B	Product C	Productive capacity, mach.-hr. per week	Total
a. Profit, $/unit	9	6	5		
b. Production rate in hours/unit					
Casting	12	6	4	200	
Machining	6	18	30	600	
Finishing	8	28	4	400	
Problem solution					
c. Optimum product quantities, units/week	7.4	10.4	12.3		
d. Profit, $/week $= a \times c$	66.60	62.40	61.50		190.50
e. Production, hr/week, used by optimum product quantities $= b \times c$					
Casting	89	62	49		200
Machining	44	187	369		600
Finishing	59	292	49		400

Notes

1. Total profit is obtained by adding the individual profits in line d.
2. Comparison of proposed production schedule (lines e) with productive capacity (lines b), shows that the program would use up all of the presently available capacity.
3. The optimum profit would result if the quantities sold of products A, B and C are in reverse magnitude to their unit profits. (Compare lines a and c.)

4. Table 6-1, enlarged to handle a considerable number of products and processes, can readily be prepared with the aid of a desk calculator. Table 6-2 could not be reasonably obtained by any than electronic data handling systems when many products and processes are involved.
5. A computer can be programmed to yield the data given by both Tables 6-1 and 6-2 separately.

There are a few, isolated instances in which a firm has decided that it desires the type of information revealed by both tables, and it usually requires only a few extra minutes on a computer to supply the Ratio-Analysis values in addition to the exact MP results. The decision to provide both types of data is usually based on the preferences of the people who must work with them.

Other Short-Cuts

Aside from the approximate ratio-analysis, we also have among the kit-bag of tools a rapid and exact procedure known as the Distribution Method. Originated for problems dealing with distribution and warehousing, it has been extended to multitudes of other areas, but nevertheless its applicability is limited to only certain types of problems. Moreover, whenever the Distribution Method can be applied, it is equally valid to use the more universal Simplex series of matrix steps in Chap. 12.

Before presenting the Distribution Method, a programming-planning solution of an inventory problem will be discussed. From the analysis of this material, the reader will gain an intuitive insight of the more formal Distribution Method procedures given in the subsequent chapter.

Summary

A rapid and approximate desk-calculator method has been presented. Its prime use is not so much to provide a ready technique when one does not wish to go to a computer, but rather to demonstrate through various ratio transformations the interlocking nature of MP problems. While the ratios do serve to disentangle and lay bare some of the key relationships among profits, production rates, and productive capacities, they usually will lead only to an approximate solution of the problem of finding an optimal allocation of resources or determining an optimal goal to pursue.

7

INVENTORY PLANNING

The application of mathematical programming need not be confined to the coordination of productive capacity with sales potentials so as to achieve the best utilization of production and sales activities. Further gains can be made if MP is extended to include the scheduling of inventory levels. This is especially true where production and sales requirements depend to a large extent on irregular demand, such as with highly seasonal items. Then the additional planning flexibilities afforded by the proper scheduling of inventory can be included in management's mathematical programming arsenal.

Illustration

The manner in which mathematical programming of inventories is accomplished profitably can be demonstrated by the following example.*

A manufacturer of products used primarily in the summer months is facing a highly seasonal demand pattern. His plant can normally produce up to 100 units each month, but he expects the following demands:

Month	Demand, units
April	50
May	150
June	300

Looking at cost factors, he sees that it will cost $10 to produce one unit in the month in which it is sold and delivered to

---

*It will be recognized that, for the sake of illustration, only a miniature example using simplified data can be used.

the customer. If the unit is produced one month early, for expected sale in the subsequent month, inventory costs raise the
unit cost to $20. This cost estimate includes all of the expenses
of warehousing, interest on borrowed inventory investment,
extra handling, and other factors involved in storage. Similarly,
production of a unit two months ahead of actual sale and delivery
raises the cost per unit to $30.

Alternatives

Aside from inventory buildup during the slack season, two
other possibilities are open to management.

First, because of the great demand in June, the company can
make arrangements to run the plant on an overtime basis, which
involves payment of time and a half for overtime wages. Also,
as a result of such heavy production schedules, extra expenses
for quality losses and waste and scrap will be incurred. Therefore, while in any month 20 additional units can be produced
through overtime—above and beyond the normal capacity of
100—the cost per unit is estimated at $15.

Second, the company can call on another firm, an outside
supplier, to produce some of the units. This firm can supply up
to 1,000 units during the months of May and June inclusive.
During May the outside supplier will charge $14 per unit, but in
June the price rises to $26.

The types of data just given can best be surveyed if placed
in tabular form, as shown in Table 7-1. On the far right-hand
side we note the market demands expected, representing 50,
150, and 300 units for April, May, and June. Productive capacity
is shown along the bottom line, with 100 units per month from
April to June, and a potential of 20 further units from June overtime and 1,000 units from outside supplier's source. The remaining large boxes show one possible production schedule
which would supply market demand and yet stay within capacity
(using the extended definition of "capacity" to include overtime plus outside supply available).

Cost per unit is given in the small squares at the upper left-
hand corner of each box. These cost data help make an initial
determination of quantities to produce in each month. For example, in April the column shows that production in that month
is least costly, at $10 per unit. Capacity of 100 units in April,
moreover, is ample to supply market demand of only 50 units.
In May, on the other hand, 150 units are needed, while capacity
is only 100. The schedule therefore calls for 100 units to be
produced in May at $10 per unit, with another 50 units purchased
from the outside at a supplier's price of $14 per unit.

An alternative would have been to produce, in April, 50 extra units for use in May, but as the intersection of the April column with May row shows, these 50 units would have cost $20 each, as against purchase for $14 from the outside. It can also be observed that it would have been entirely feasible to have scheduled 50 units for April production and May use, because the total April capacity of 100 units would then have been just filled up. (50 units made and used in April plus 50 units made in April for use in May.)

Table 7-1 Market Demands, Productive Capacity and Inventory Costs

Month in which demand occurs	Month in which production occurs			June output at overtime	Purchase from outside supplier	Market demand
	April	May	June			
April	10 50					50
May	20	10 100			14 50	150
June	30	20	10 100	15 20	26 180	300
Productive capacity	100	100	100	20	1000	

Shaded boxes represent impossible solutions. For example, we cannot in May produce for sale and delivery in April.

Evaluation of Schedule

The initial schedule may now be evaluated as to costs. Later on, subsequent schedules can be tested to note whether or not an improvement was attained.

Examining all boxes for which production is scheduled or during which purchase from a supplier is planned, we find the results tabulated in Table 7-2.

The cost shown is not necessarily the best one, since a great many alternatives were not investigated, such as the possibility of producing in May and April for June. At first glance, such a schedule of production to inventory may look uneconomical, but this may not necessarily be true from an overall viewpoint. Also, while the present highly simplified problem may reveal some desirable alternatives "on sight," such is not at all to be expected without mathematical aids in a really large-scale problem. In place of the six columns of alternatives shown, a real problem involving weekly instead of monthly breakdowns and a large number of products might cover some 1,000 or more columns.

Table 7-2 Cost Evaluation of Initial Schedule

(a) Month of production or purchase	(b) Month in which use is scheduled	(c) Number of units scheduled	(d) Cost per unit $	(e) Total cost = (c) × (d) $
April	April	50	10	500
May	May	100	10	1,000
May	May	50	14	700
June	June	100	10	1,000
June	June	20	15	300
June	June	180	26	4,680
			Total, overall	8,180

Optimum Solution

Application of mathematical programming principles leads to the solution shown in Table 7-3. Because of the relatively small size of the problem, the computer was able to find only one improvement in the original schedule. This improvement is as follows:

Although the original schedule called for production of 100 units in May for consumption in May, *do not produce any units in May*. Instead, *increase the purchase from supplier by 100 units to 150* as a total. True enough, the 100 additional units bought from the supplier will cost $4 more per unit (compare $10 per unit of production against $14 per unit of purchase). But, by purchasing these 100 units, we are free *in May to produce 100 units to inventory, for use in June*. Next, we can *purchase 100 units less in June* from the supplier. The cost saving per unit is $6 (compare $20 for the inventoried as against the purchase cost in June). This $6 saving in June outweighs the $4 extra cost in May, and the total net saving is thus: 100 units × ($6 − $4) = 100 units × $2 = $200.

Table 7-3 Optimum Solution

Month in which demand occurs	Month in which production occurs			June output at overtime	Purchase from outside supplier	Market demand
	April	May	June			
April	10 \ 50					50
May	20 \	10 \			14 \ 150	150
June	30 \ 100	20 \ 100	10 \ 20	15 \	26 \ 80	300
Productive capacity	100	100	100	20	1000	

The new total cost is the cost of the old schedule, $8180, reduced by $200, giving $7980.

Further Applications

Coordination of sales and production, as we have seen, can be strengthened further through mathematical programming, by including an analysis of inventory factors. Although the particular illustration related to inventory of finished goods, it is apparent that parallel principles are also applicable to the following:

1. Inventory of semifinished goods, products, or hardware to be assembled, finished, and completed upon order at a later time.
2. Inventory of supply or raw materials, in the expectation of later shortages.
3. Desirability of advance commitments regarding purchases from suppliers, if special discounts are thus obtainable.
4. Need to build up labor force, especially in the skilled and semiskilled categories, to anticipate future needs.
5. Justification of equipment or machinery purchase or rental, based on the mathematical programming evaluation of alternative costs.

Summary

Planning of inventories, as we have seen, can be aided greatly through Mathematical Programming analysis. While many approaches to inventory scheduling can be used, a systematic manner is preferable not only from a viewpoint of orderly planning procedures but also because such a step-by-step method is more likely to lead us to a solution that will prove optimal in practical applications.

Inventory planning problems of the type just discussed involve a mathematical analysis procedure labeled "The Distribution Method of Mathematical Programming," because the steps of solving it were first developed in connection with distribution problems. An application in this area will be presented next.

8

DISTRIBUTION METHODS

When problems of optimal distribution systems need solving, a special set of techniques known as Distribution or Transportation Methods, are very useful. Several approaches have been formulated. We will discuss the most common one, known as the "stepping stone" technique, because it utilizes a step-by-step analysis of feasible distribution assignments until an optimal program has been found.

Illustration

A feed mix manufacturer has three plants, located at Alton, Belton, and Carlton. At each of these towns, there is within-plant storage to supply local demand. Capacity of each plant exceeds these needs, and the overage is available for shipment to three distributing warehouses at Denville, Eastville, and Fernville. Weekly tonnages required at the warehouses are 4, 6, and 18, respectively, based upon the sales forecast for the winter season. Excess tonnages available are 6, 10, and 12 at Alton, Belton, and Carlton, respectively. Transportation costs vary, as shown in terms of the dollars-per-ton figures in Fig. 8-1.

The question now is this: What shipping schedule will fulfill the tonnage requirements from anticipated availabilities while at the same time minimizing transportation costs? A small-scale problem could be readily solved by trial-and-error, a really large distribution system with many and widely dispersed outlets calls for a Mathematical Programming approach.

Initial Assignment

Begin by developing an initial assignment of shipments from the various locations to the warehouses. The assignment must

To From	Denville warehouse	Eastville warehouse	Fernville warehouse	Tonnage available
Alton plant	$1	$2	$3	6
Belton plant	$1	$1	$4	10
Carlton plant	$2	$5	$2	12
Tonnage required	4	6	18	

FIG. 8-1 Distribution problem. Three warehouses at the locations shown require tonnages of feeds produced in three other locations. The shipping costs, in pounds per ton, are given in the boxed inserts for each of the 9 "from–to" squares.

be in conformity with the so-called "rim conditions," that is, the amounts available and required, as shown in the bottom row and far right-hand column "rimming" the matrix in Fig. 8-2.

In order to keep track of successive assignments, it is best to follow a systematic procedure in loading each of the squares. Traditionally the northwest corner rule is followed, which starts at the intersection of the Alton and Denville assignment (cell AD) moving downward and to the right, until all cells are filled. Observe the following:

1. Upon loading AD with the 4 tons required by Denville, we have taken care of all the needs of this warehouse and move on to the Eastville column.
2. Since 2 out of Alton's 6 weekly tons are not yet assigned, we allocate them to AE. But Eastville needs a total of 6 tons, so we move downward to Belton for the remaining 4 tons to be loaded to cell EB.
3. Finally, of Belton's 10 tons, the remaining 6 are given to Fernville. A further 12 tons from Carlton to Fernville will both exhaust Carlton and fill all of Fernville's needs.

The program has matched available capacities with anticipated requirements. But is it optimal? As a step toward ascer-

To From	Denville (D)	Eastville (E)	Fernville (F)	Available
Alton (A)	$1 (4)	$2 (2)	$3	6
Belton (B)	$1	$1 (4)	$4 (6)	10
Carlton (C)	$2	$5	$2 (12)	12
Required	4	6	18	

FIG. 8-2 Initial assignment of shipments for distribution problem. Beginning with the northwest corner and proceeding downward and to the right ("northwest corner rule") we first ship the required 4 tons to Denville from Alton. But since Alton produces 6 tons a week, 2 tons are left over to ship to Eastville. The remaining 4 tons needed by Eastville are now supplied from Belton, reducing its capacity to 6, which now goes to Fernville. All of Carlton's capacity is required to supply Fernville with an additional quantity of 12 tons to make up the total of 18 tons needed.

taining the answer to this question, let us first calculate the cost of the program.*

Cost Evaluation

Analysis of the effects of the initial distribution schedule, in Fig. 8-3, yields a total weekly cost of $64. In order to investigate whether a lower cost is achievable, we may now study the per-ton effects of movements from loaded to unloaded cells, keeping in mind the rim conditions. Proceed column by column and downward within each column. Moreover, in order to automatically satisfy the rim conditions, a rectangular set of moves among adjoining or at least nearest-by cells, is required. This set of rules is best explained by an example.

---

*Since our program has resulted from the systematic stepping from one cell to the next in making assignments, the procedure is also known as the "stepping-stone" method.

First to be evaluated is the open cell BD. A shift of assignment from BE to BD means $2 per ton less in BE and $1 per ton more in BD, thus − $2 + $1, or a net saving of $1 per ton. However, the movement made must be balanced by another movement to restore the requirements of the rim condition. Therefore, moving a ton from AD to AE, to counterbalance the shifts in the Belton row, results in − $1 + $2, or a net of $1 per ton more. The savings are thus balanced by the losses, and no net gain is achieved by a shift to BD.

To ⁄ From	Denville (D)	Eastville (E)	Fernville (F)	Available
Alton (A)	$1 (4) $4	$2 (2) $4	$3	6 — $8
Belton (B)	$1	$2 (4) $8	$4 (6) $24	10 — $32
Carlton (C)	$2	$5	$2 (12) $24	12 — $24
Required	4 — $4	6 — $12	18 — $48	Total cost $64

FIG. 8-3 Cost of initial assignment. In each square, the dollar cost per ton is multiplied by the number of tons to be shipped, giving total cost of the assignment. For example, shipment from Alton to Eastville costs $2 per ton, which is multiplied by the 2 tons to be shipped, giving a total cost of $4. Cross-totals (entered in "Available" column) and column-totals (entered under "Required") add up to an overall cost of $64.

Next we ought to evaluate CD, but we run into difficulties because there is no rectangular shifting possible. Had there been an assignment in AF, then we could have checked these shifts: CF to CD and, counterbalancing, AD to AF. For situations of this nature, we may follow a modified route, proceeding either clockwise or counterclockwise, making sure that we are stepping only among loaded cells and keeping our turns at right angles. Figure 8-4 shows the evaluation. It indicates that a movement to load CD would result in increased costs per ton, so we will ignore this possibility.

To From	Denville (D)	Eastville (E)	Fernville (F)	Available
Alton (A)	– \$1 ④ ⟶	+ \$2 ②	\$3	6
Belton (B)	\$1	– \$2 ④ ⟶	+ \$4 ⑥	10
Carlton (C)	\$2 + ⟵	\$5	– \$2 ⑫	12
Required	4	6	18	

FIG. 8-4   Evaluation of empty square CD. Square CD refers to shipment from Carlton (C) to Denville (D). If we were to supply one of the required 4 tons of Denville from Carlton (+ for CD) we would have to take it from Fernville (– for CF). But, by adding to CD, we now must remove one ton from AD (– for AD) or else Denville would be receiving more than 4 tons. However, with 1 less in the Alton row, we must now add 1 ton to the AE square. This process continues in the pattern shown.

Note that the pattern must always represent a closed path, starting at the square evaluated, with right angle turns only at squares that already have tonnage assigned to them. Clockwise or counterclockwise loops are permissible.

Evaluation above leads to: + \$2 – \$1 + \$2 – \$2 + \$4 – \$2, by evaluating the path shown, beginning with the CD square and showing the dollar cost of each ton added or removed from a square. The net result of these additions and subtractions is + \$8 – \$5 = \$3 additional cost for moving tons as indicated. Therefore, by leaving the CD cell empty, we are better off.

Loading of cell CE calls for a movement from CF, counterbalanced by a shifting from BE to BF. The effects are – \$2 + \$5 – \$2 + \$4 or a net increase of \$9 per ton, so that this square is best left empty. We do, however, realize a gain by loading the last remaining empty square, FA. Shifting AE to AF with a counter of BF to BE involves these per-ton cost changes: – \$2 + \$3 – \$4 + \$2 or a net saving of \$1 for each such ton.

New Assignment

Taking advantage of the potential cost saving, we develop a revised assignment in Fig. 8-5. Two tons were shifted from AE to AF, leaving 0 tons in the former cell. Correspondingly, 2 tons were moved from BF to BE, so that the latter has now 6 and the former 4 tons. The movement of 2 tons is the maximum amount possible within the limitations imposed by the rim conditions.

To From	Denville (D)	Eastville (E)	Fernville (F)	Available
Alton (A)	$1 ④	$2	$3 ②	6
Belton (B)	$1	$2 ⑥	$4 ④	10
Carlton (C)	$2	$5	$2 ⑫	12
Required	4	6	18	

FIG. 8-5   Revised assignment. The 2 tons of AE have been moved to AF. As a result, we now need to move only 4 (instead of the prior 6) tons from Belton to Fernville. In turn, BE now receives 6 instead of the prior 4 tons. All these changes are consistent with the amounts in the available and required sections (also known as "rim conditions").

The net effect, per ton, of the changes is to move tonnage out of the expensive ($4 per ton) BF cell. The other cells affected, AE, AF, and BE have costs lower per ton.

A new evaluation of all the cells is now made, which shows that no further improvements are possible and that the revised assignment is an optimal one. It is, however, of interest that the evaluation of square BD yields neither an improvement nor a loss. In particular, we have − $2 + $1 − $1 + $2 = 0. Therefore, as an alternative optimum schedule, we would have a program similar to Fig. 8-5 excepting that in the upper left-hand segment, the assignments are: AD = 0, AE = 4, BD = 4, and BE = 2 tons.

The revised schedule results in a total cost of $62, as calculated in Fig. 8-6. Although the saving of $2 per week against the initial cost of $64 is small, it must be realized that in really large-scale problems the benefits can be very considerable.

From \ To	Denville (D)	Eastville (E)	Fernville (F)	Available
Alton (A)	$1 ④ $4	$2 $0	$3 ② $6	6 $10
Belton (B)	$1	$2 ⑥ $12	$4 ④ $16	10 $28
Carlton (C)	$2	$5	$2 ⑫ $24	12 $24
Required	4 $4	6 $12	18 $46	$62

FIG. 8-6   Costs of revised assignment. Total cost of $62 is below the original cost of $64.

Optimum Assignment

Although an improvement has been achieved, the lowest cost has not as yet been attained. One further revision of the shipping schedule, as shown in Fig. 8-7, is required. In effect, this shift in schedule utilizes the fact that shipment to Fernville is less costly from Alton (at $3 per ton) than from Belton (at $4 per ton). A total of four tons can be shifted in this manner, resulting in a saving of $4. Total cost of the program is now $58, as compared to the original $64. A check of this latest program, moreover, will reveal that no further improvements are possible. An optimum in terms of lowest transportation costs has been attained for this distribution problem.

From \ To	Denville (D)	Eastville (E)	Fernville (F)	Available
Alton (A)	$1	$2	$3 ⑥	6
			$18	$18
Belton (B)	$1 ④	$2 ⑥	$4	10
	$4	$12		$16
Carlton (C)	$2	$5	$2 ⑫	12
			$24	$24
Required	4	6	18	
	$4	$12	$42	$58

FIG. 8-7 Final assignment, with lowest cost of $58. This is considerably below the original cost of $64.

Demand and Capacity Comparison

Demand balanced available supply in the illustration above. Suppose, however, that Fernville's demand had been only 10 instead of 18 tons. The excess of 8 tons can be readily handled by creating a nonexistent or so-called "dummy" warehouse that demands these 8 tons. Since tons are shipped only in imaginary terms, the cost per ton is zero. The optimal solution then assigns all but the 8 tons to real warehouses in the most economical distribution pattern and also loads the dummy warehouse with the 8 spare tons. No actual production, shipment, or storage will, of course, take place for the dummy.

A Caution on Zero Loads

In making evaluations for possible cost improvements, utilizing rectangular patterns or their modifications, it was emphasized that we can step only from a loaded cell. In some instances, such as for evaluating BD of the matrix in Fig. 8-5, we utilized AE with zero load. Only one such zero load may be used in an evaluation of a rectangle. We may, however, skip over an empty cell, as was done, for example, in Fig. 8-4 for squares CE and BD.

General Applicability

Not just distribution problems, but many other situations calling for the assignment of available resources to anticipated demands, are amenable to the solution procedures just outlined. For example, a large-scale development program involves several tasks (demands) that can be handled by several engineers and scientists on the staff (capacities). However, the engineers and scientists have different educational, experience, and related background, making them more or less qualified for each major task. The problem then resolves itself into a Distribution Method approach, whereby the overall optimum assignment of men to tasks is accomplished.

In planning production schedules, keeping in mind anticipated monthly requirements, productive capacities, inventory carrying costs, and overtime expenses, the Distribution Method helps us find an optimal schedule. An example was given in Chap. 7.

A wide range of assignment problems can thus be treated by means of the distribution matrix, and the calculation steps will be generally less demanding than in the basic matrix procedures.* However, the distribution method is applicable only when capacities and requirements can be expressed in like units (tons in our example) and there are no interlocking restrictions (for example, output of one plant is not dependent on that of another plant or shipment is not serial in one truck to two warehouses, each receiving a partial load). When the distribution method is inapplicable, resort to the more universally applicable basic matrix procedures is then taken.

*Known also as the Simplex Method, these basic matrix procedures are given in Chap. 14.

9

OPTIMAL ASSIGNMENTS

A general area in which MP has further considerable usefulness is in the scheduling of assignments to minimize costs or maximize results. An example will illustrate this application.

Illustrative Example

In the production of yarn filaments, failure of the pumps metering out the liquid polymer is a relatively costly item. It was found that pump failure rates varied with the particular product style produced, since each style utilized varying rates and amounts of liquid polymer or "dope" as it is called in the industry.

A study of past failure rates at one filament extrusion plant indicated that a further factor in failure rate occurrence was the particular type of dope metering pump (Type I, II, or III) utilized with any one of the 10 yarn styles (A to J). This information, together with the season's market demand for each style, is provided in Columns (1) to (3) of Table 9-1.

The question to be solved by MP was: "How do we assign pumps to each style so as to minimize total pump failures?" Application of the Distribution Method leads to the recommended assignment (column 4), which in turn involves some 1,481 failures. At a cost of $10 per pump, this represents a cost of $14,810. Utilization of the recommended assignments, to the extent possible, in actual plant operations resulted in approximately this cost. It was 20% below the level that had prevailed previously, when pump utilization was based on unaided judgment as the sole guide.*

*This case history is based on a paper by J. W. Cowdery, An Application of Linear Programming to Dope Metering Pump Replacement, presented before the Statistics Section, Virginia Academy of Science, Roanoke, May 8, 1958.

Table 9-1 Dope Metering Pump Assignment

Problem data

(1)	(2)	(3)		
		Failures per 1000 hr of operation of pumps		
Product style	Production required, machine-weeks*	Type I	Type II	Type III
A	153	3.36	–	0.70
B	33	0.84	–	2.10
C	24	2.87	–	0.70
D	14	6.72	–	0.77
E	98	11.34	5.81	6.37
F	63	1.68	0.21	0.77
G	3	0.91	–	0.91
H	36	1.47	0.84	0.98
I	50	1.12	0.70	–
J	110	4.76	3.08	4.34
Pumps available, no.		338	97	194
				Totals

Notes: "–" means "no failure data available."
* Machine-weeks required = No. of pumps required.
† Note that 45 pumps of Type I are not used.

For the three different types of pumps (I, II, III), the failure rates vary with the product style (A–J) extruded. The production required to meet current demand is shown in terms of machine weeks. Product styles (1), production requirements (2), and failure rates (3), thus represent the *Problem data*, in the sense that product styles must be assigned to the pumps so as to meet pro-

Application to Human Factors

An interesting application of MP to human factors occurs in the assignment of personnel of various proficiencies to the demands of several tasks. The delegation of tasks to each person must be in such a way that overall proficiency is maximized. For this purpose, proficiency indexes can be assigned to each candidate, measuring his relative capability to cope with a variety of tasks. The indexes, in turn, are based on personnel tests. Correspondingly, the tasks are marked as to the degree of skill and other qualifications required. Thus, for three men and three tasks, there would be 3 × 2 or 6 possible combinations of assignments, and MP selects the best of these. For a

Problem and the Optimum Solution

	Minimum-failures solution			
(4) Optimal assignment of no. of pumps to styles			(5) Expected failures per 1000 hr of plant operation	
Type I	Type II	Type III	$(3) \times (4) = (5)$	
—	—	153	$0.70 \times 153 =$	107.10
33	—	—	$0.84 \times 33 =$	27.72
—	—	24	$0.70 \times 24 =$	16.80
—	—	14	$0.77 \times 14 =$	10.78
—	95	3	$5.81 \times 95 + 6.37 \times 3 =$	571.06
63	—	—	$1.68 \times 63 =$	105.84
3	—	—	$0.91 \times 3 =$	2.73
36	—	—	$1.47 \times 36 =$	52.92
50	—	—	$1.12 \times 50 =$	56.00
108	2	—	$4.76 \times 108 + 3.08 \times 2 =$	520.24
293†	97	194		1481.19

duction requirements while at the same time minimizing total failures. The optimal assignment for this purpose is shown in column 4, which leads to a total of 1481 failures per 1,000 hr of plant operation, or roughly 1½ pumps per hour. In practice, it is not possible to make assignments exactly corresponding to the optimum, but the schedule given in column 4 serves as a guide. The extent to which actual pump failures in yarn extrusion operations exceed the minimum predicted of 1,481 in column 5 represents the success achieved in relation to the goal.

large-scale problem of, say, eight tasks and men, there would be $8 \times 7 \times 6 \times 5 \times 4 \times 3 \times 2 = 40{,}320$ possible assignments to choose from for the optimal combination.*

Allocation of Salesmen

The size, type, and nature of potential customers may vary among different sales territories. Salesmen, in turn, have dif-

*See, for example, Paul A. Young, On the Application of Linear Programming Techniques to Human Factors in Space Programs, paper in "Proceedings of the 7th Military-Industry Missile and Space Reliability Symposium," North Island Naval Air Station, San Diego, Calif., June, 1962.

fering qualifications as regards their ability to sell to various types of customers. Differences may be in terms of size of stores, merchandise mix carried, and general business conditions within localities. Some salesmen may be more adept at selling to large houses, others to small stores; moreover, their experience with various types of items in the firm's merchandise line may be different. Salesmen's qualifications can be catalogued and sales territory requirements analyzed. Next, a Distribution Method solution of the problem will indicate what allocation of sales territories to individual salesmen is likely to lead to highest overall sales.

Loom Assignment

A textile mill had noted that the frequency of yarn breaks during weaving varied not only with the fabric style produced but also the type of loom on which the fabric was run. Yarn breaks produced deficiencies in quality while at the same time causing the loom to automatically stop and wait until the break was repaired. Considerable reductions in downtime and enhancement in quality resulted from the use of MP to allocate various fabric styles to looms in a manner which minimized yarn breaks.

Summary

A variety of illustrations have been given in order to demonstrate the abundance of possible areas where an organization can improve operations through MP assignments. Usually, the Distribution Method is applicable.

10

PROGRAMMING FOR QUALITY
AND RELIABILITY

The safety and success of a space mission, the dependability of a ballistic missile, the durability of a car, and the life of a washing machine all depend on the quality and reliability of the components. Beyond this, however, there must be reliability design. This refers to the fact that certain components of a complex system often cannot be produced to the theoretical quality or desired reliability needed for system performance. We can, however, provide redundancy for critical components or modules, such that when one part fails there is an automatic switchover to at least one extra component. In lieu of automatic changes, when a manned flight is involved or some other mission where manual repairs are possible, a set of spare parts may be provided.

But there is a limit to the amount of redundancy or the number of spares that can be included, because total volume, weight, and cost of a system must stay within engineering, space, and budgetary bounds. It is in this situation where MP evaluates the degree of redundancy or number of spares for each component or module that will maximize overall mission reliability, system reliability, or equipment life. Redundancy need not be in terms of numbers. For example, quality and reliability can often be improved by utilizing stronger and heavier gears, pistons, and rods. But again, there is a limit in terms of weight, space, and cost.

The project engineer in charge of quality and reliability should not only be familiar with the managerial-technological aspects of his problem, but also with the manner in which the computer must be fed the pertinent variables in identifiable terms. In this way, he and the computer programmer can communicate in a cooperative endeavor toward problem solutions for complex systems reliability attainment.

An Illustration

In order to demonstrate how MP works, we may employ an illustrative example adapted and simplified from a problem of providing an adequate provision of spare components to assure system reliability. Since these spares were to be carried on an airborne mission, it was essential to take account of weight and space limitations. Although the real situation may involve many dozens of types of modules and components and a variety of limiting factors, we will confine ourselves to an examination of a case with only two components, designated as Type I and Type II.

Reliability tests had indicated an expected failure rate of 0.8 per mission for Type I and 1.5 per mission for Type II. Thus, in ten missions we might typically expect to have to replace 8 components of Type I and 15 components of Type II. Although a standing rule had been established to "provide at least one spare of each critical component," it seems obvious that in many missions more than that will be needed. There were, however, volume and space limitations that made it impossible to carry more than either four components of Type I or 6 components of Type II. The details leading to this result appear in lines *a* to *e* of Table 10-1.

Plotting Results

We may plot the data, as in Fig. 10-1, resulting in the shaded area of feasible solutions. That is, within this area we can provide certain numbers of components that are consistent with all restrictions given by the facts of the problem. For example, we could provide two components each of Type I and II. But a quick look will reveal that with two components of Type I we *can* have as many as four components of Type II, and obviously this amount would assure greater spares redundancy. Points *A*, *B*, *C*, and *D* which bound the feasible area for the largest number of components, are likely to contain the optimum combination. The question is, which is the one?

Reliability Evaluation of Points

Begin by eliminating point *D* as a likely candidate, since we cannot carry 3 1/2 components. If we are to use but three spares of Type I, then we might as well utilize the now feasible two spares of II; but that gets us to Point *C*. The various

Table 10-1 Spare Parts Evaluation for Mission Reliability Problem

Problem data and analysis	(1) Component Type I	(2) Component Type II	(3) Total
a. Volume of each component, in cc	400	200	
b. Weight of each component, in grams	400	400	
c. Limitations on space available, in cc			1600
d. Allowable weight limitations, in grams			2400
e. No. of components useable if only type I *or* II but *not both* are carried as spares			
Volumetrically $= 1600/a$	4	(8)	
Weight-wise $= 2400/b$	(6)	6	
(Parentheses indicate that no. is not feasible. For example, 8 of type II cannot be taken because the weight limitation is 6).			
f. Feasible combinations of type I *and* II jointly, given by points bordering the shaded area of Fig. 10-1:			
Point A	1	5	
B	2	4	
C	3	2	
g. Probability of running out of spares (read directly from Poisson tables), in percent (rounded):			
Point A	19	0	19
B	5	1	6
C	1	12	13
(Since a run-out on either type I *or* II will cause mission failure, the two percentages have been added to obtain the failure probabilities in column 3)			
h. Reliability of mission, in percent ($= 100\% - g\%$)			
Point A $= 100 - 19$			81
B $= 100 - 6$			94
C $= 100 - 13$			87
i. Optimum components for maximum attainable reliability of 94% is, from line f, point B:	2	4	

Conclusion: To attain the maximum reliability of 94% within the volume and weight limitations of the mission, carry two spares of type I and four spares of type II.

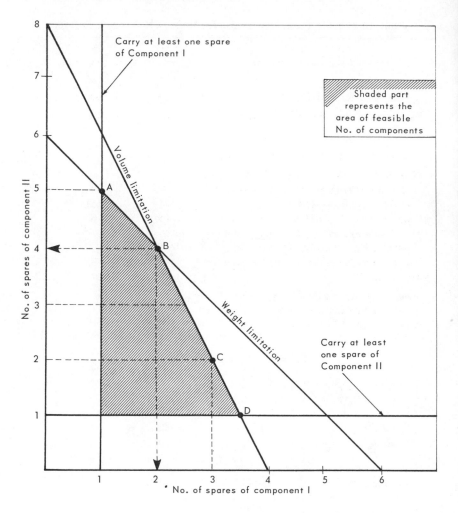

FIG. 10-1 Reliability problem analysis. Volume limitation line is found by connecting the 4 on the Component I scale with the 8 for Component II scale, with the values 4 and 8 derived in line e of Table 10-1. The weight limitation line is found similarly. The requirement that at least one spare be carried is also entered as shown. Within the limiting lines so established lies the feasible area (shaded), with points A, B, and C representing likely candidates for optimum spares combination. Point D, representing fractional components, is no better than C and need not be evaluated. Points A, B, and C are now reviewed to discover which of them leads to highest reliability.

combinations represented by *A*, *B*, and *C* are given in line *f* of Table 10-1. Thus, point *A* permits 1 of I and 5 of II, B allows 2 of I and 4 of II, and *C* yields 3 of I and 2 of II. Which combination is optimal will depend on the likelihood that we will need all of the spares. Another way of looking at this problem is to say:

Assuming we carry the spares shown, what is the likelihood that we will be caught short because the need for replacement of failing components exceeds our supply of spares during the mission?

Percentage points of the cumulative Poisson probability distribution such as in Table 10-2, will yield the required probabilities which are entered in line g of Table 10-1. From probability of failure it is but one step, by means of subtraction from 100, to arrive at reliability in percent. We note that, within the constraints of the problem data and requirements, by carrying two components of Type I and four of Type II, we can maximize reliability at 94%.

The reader will note that for further components, we would need additional dimensions to analyze; and while humans can hardly visualize more than three dimensions, mathematically even 50 dimensions pose no problem for the computer.

Extended Formulation

In practice, we may have to deal with several critical components for an individual subsystem, which in turn is part of a large and complex total system. Overall system reliability then becomes relatively difficult to evaluate and the designing for ultimate quality and reliability poses numerous problems.

It is apparent that we will no longer be able to apply the graphic methods, which are limited to two or at most three critical components. Moreover, we must find a way to combine subsystem reliability factors with total system effects.

The Simplex series of matrix steps, to be presented in a later chapter, are helpful in these evaluations. But even here we may face limitations. In particular, when in a system each subsystem contains a large number of component types, the evaluation of points may overtax the computer, and resort to the special techniques of Dynamic Programming is required.*

Further Applications

Mathematical Programming problems tend to repeat themselves in various guises, but the basic structures are unvarying. A goal or objective is desired, such as maximizing reliability,

*R. Bellman, "Dynamic Programming," Princeton University Press, Princeton, N. J., 1957.

Table 10-2 **Probabilities of Component Failures**

Probabilities that the number of failures or more below will actually occur in any one particular mission

Probabilities in percent

(Blank spaces indicate a probability of nil)

Average failures per mission	0	1	2	3	4	5	6	7	8	9	10	11	12	13	14
0.01	100	0.1													
0.02	100	2.0													
0.03	100	3.0													
0.04	100	3.9													
0.05	100	4.9	0.1												
0.06	100	5.8	0.2												
0.08	100	7.7	0.3												
0.10	100	9.5	0.5												
0.12	100	11.3	0.7												
0.15	100	13.9	1.0												
0.20	100	18.1	1.8	0.1											
0.40	100	33.0	6.2	0.7											
0.60	100	45.1	12.2	2.3	0.3										
0.80	100	55.1	19.1	4.7	0.9	0.1									
1.00	100	63.2	26.4	8.0	1.9	0.4									
1.20	100	69.9	33.7	12.1	3.4	0.8	0.1								
1.50	100	77.7	44.2	19.1	6.6	1.9	0.4	0.1							
1.80	100	83.5	53.7	26.9	10.9	3.6	1.0	0.3							
2.00	100	86.5	59.4	32.3	14.3	5.3	1.7	0.5	0.1						

2.50	100	91.8	71.3	45.6	24.2	10.9	4.2	1.4	0.4	0.1					
3.00	100	95.0	80.1	57.7	35.3	18.5	8.4	3.4	1.2	0.4	0.1				
4.00	100	98.2	90.8	76.2	56.7	37.1	21.5	11.1	5.1	2.1	0.8	0.3	0.1		
6.00	100	99.8	98.3	93.8	84.9	71.5	55.4	39.4	25.6	15.3	8.4	4.3	2.0	0.9	0.4
8.00	100	99.9	99.7	98.6	95.8	90.0	80.9	68.7	54.7	40.7	28.3	18.4	11.2	6.4	3.4
10.00	100	99.9	99.9	99.7	99.0	97.1	93.3	87.0	78.0	66.7	54.2	41.7	30.3	20.8	13.6

Example: A component is expected, based on reliability tests, to exhibit an average of 0.8 failures per mission (defined in terms of time length and severity of environmental stresses). Then we may expect these failures: Zero or more, 100% of time, one or more, 55.1 and two or more 19.1% of the time. Therefore, if only one spare is carried on a mission, then 19.1% of the time the failures will exceed the available replacement spares.

Source: Failure probabilities were derived from the assumption, supported on both theoretical and practical grounds, of a so-called Poisson distribution, expressed mathematically as $e^{-m} \times m^i/i!$. Here e is the well-known base of the natural logarithm, m represents the average failures per mission and i the expected failures in any particular mission. Values of the Poisson distribution have been published widely, the most useful reference being E. C. Molina, Poisson's Exponential Binomial Limit, Van Nostrand, Princeton, N. J., 1941. Readers are referred to this (or other) tables for those instances when wider or more detailed values than those given here are needed. When using such other sources, individual Poisson values must be cumulated to obtain the type of probabilities shown above.

minimizing costs, or attaining optimal process yield. In turn, the road toward accomplishing these goals by various alternative combinations of factors, is constrained by limitations. Considering these limitations, we must find the optimal solution. Some further applications will highlight this point.*

Minimizing Quality Control Costs. A company has several classes of inspectors with varying speed and accuracy. Production schedules call for a daily or weekly minimum number of parts to be inspected. Varying speeds and accuracies of inspectors result in different inspection costs while the lack of accuracy can be measured in terms of dollar losses from the passing of off-standard goods; viz., the cost of inspection error. Mathematical Programming then seeks to attain the called-for amount of inspection at minimum cost.

Reliability Apportionment. A system, based on two subsystems in series, must attain a reliability of 90%. Initial evaluation of the two subsystems reveals reliabilities of 85 and 87% for subsystems I and II, respectively, with resultant overall reliability of 85 × 87 or 74%. Subsystem reliabilities must thus be improved to attain the 90% goal. The relative additional program cost for incremental reliability improvement is determined to be in a ratio of 0.3 to 0.7 for subsystems I and II, respectively, with a corresponding reliability improvement tradeoff factor of 0.9 and 0.1. The problem is to minimize the costs to meet the 90% reliability requirement.

Failure Rate Apportionment. The times to failure of a serial system are exponentially distributed, consisting of several components at various costs and failure rates. Given a maximum failure rate and limitations on the total amount of redundancy, develop a failure rate apportionment that will minimize failure costs.

Quality-Related Applications

There are many applications that are related to quality. For example, firms rely on MP to determine the optimal blending of components—gasolines, feedstuffs, alcoholic spirits, foodstuff

---

*These examples are due to Victor Selman, Systems Sciences Corp., Falls Church, Va., and Nelson T. Grisamore, Assistant Dean of Research, George Washington, University, Washington, D. C., as published in the "Proceedings of the 1966 Annual Symposium on Reliability," pp. 696-703, San Francisco, Calif., January, 1966.

ingredients, etc.—from a viewpoint of satisfying a given set of quality standards with combination of ingredients that will minimize costs.

A particularly useful and interesting MP case history is due to J. W. Cowdery in the replacement of dope-metering pumps in a plastics-extrusion operation as to minimize pump failures and repair costs as given in Chap. 9. The application was based on the observation that various types of pumps tended to have different failure rates, depending on the particular type of material extruded. Thus, by proper distribution of pumps to materials, minimum failures were assured.

Summary

Mathematical Programming, an outgrowth of modern management science developments, is beginning to have direct value in the quality-reliability assurative functions.* This chapter has sought to document the directions in which applications of this technique can be profitably made, provided quality-reliability engineers and managers will properly recognize and utilize this tool when and where needed.

---

*Other management science techniques used are statistical sampling plans for quality and reliability, control charting and tolerance analysis. See, for example, N. L. Enrick, "Quality Control and Reliability," Industrial Press, New York, N. Y., 1966, and E. L. Grant, "Statistical Quality Control," McGraw-Hill Book Company, 1964.

11

PROGRAMMING AND SYSTEMS MANAGEMENT

Mathematical programming (MP) applied to the coordination of the sales and production functions of an organization represents an instance of an integrated management system. MP is part of the field of Operations Research and Management Science, in its turn a technique of systems analysis and design.

Systems analysis includes in its scope not merely the study of existing systems and their various facets, but also the evaluation of alternative future systems. Obviously, the number of variables to be dealt with in such a study is large, the risks and uncertainties to be taken into account are manifold, and the problem of deciding on long-range goals or what they ought to be may be as difficult as the delineation of realistic roads to their attainment. Thus, while the term systems analysis may suggest tidy, neat, and clear-cut procedures, in fact the path toward a good ultimate design may be untidy, thorny, and at times frustrating.

Definition of Systems

One of the best definitions of systems analysis was developed in a military context, but the words apply with equal force to business systems*:

A study of the process in which costs and risks of alternative patterns of resource allocation are systematically examined and balanced against expected benefits, with the

*Air Force Manual, no. 300-4, vol. 2, part II, p. 22.

end objective of improving the quality of decision-making.

The Systems Engineering Committee of the American Society for Quality control defines a system as:

A group of interacting human and/or machine elements, directed by information, which operate on and/or direct material, information, energy and/or humans to achieve a common specific purpose.

Systems Today and Yesterday

Since systems have always been with us, we should compare modern management systems concepts with traditional ones. Table 11-1 makes such a comparison, which is admittedly biased against the "old ways," because of the author's enthusiasm for the potential and successes of newer approaches. It is not intended that the comparison imply any criticism of earlier methods. The simple fact is that most of the management technologies that buttress modern systems work have come into being only recently. Progress in the development of a valuable kit-bag of systems-oriented methodologies has undoubtedly been spurred by the availability of high-speed computers. These permit us to perform the multitudinous analysis calculations for systems evaluation and design which would otherwise not be feasible.

In these times of rapidly developing technologies, both engineering and management systems can be already obsolescent while still in an early stage of use. Nevertheless we cannot escape the ancient necessity for choice arising out of the scarcity of available resources. And among these scarcities we must list the number of qualified systems analysts.

Systems Management

Systems may be viewed as intertwined with what are rapidly becoming recognized as Modern Management and Modern Engineering. In fact, many operational aspects can be cited to demonstrate how systems methodology links these two functions, such as outlined in Fig. 11-1. In turn, the management of systems is conducted as a series of decision and control processes within an integrated network of men, materials, equipment, and information flows, involving continuous feedback of data on the basis of which the system may be further improved and adapted in response to dynamic changes of business conditions (see Fig. 11-2).

Table 11-1. Management Systems Yesterday and Today

	YESTERDAY (Conventional)	TODAY (New Era)
OBJECTIVE	Provide automatic routines, review results, make periodic revisions, based on tactics-oriented goals.	Develop procedures based on all facets of a systems complex. Strategic aspects are highlighted by showing management how key variables interact and what principal factors dominate the alternatives.
NATURE	Conglomeration of sub-systems, often involving a fragmented "bits-and-pieces" approach, with limited co-ordination of organizational elements.	Aim towards a "totally planned" system, engineered so as to integrate the dynamic relationships among the interdependent components of an organization.
GROWTH	New system superimposed on old structure, with occasional "bursts" of systems work in response to crises. The system may thus be "jerry-built".	Designed on the basis and in support of periodic long-term planning, utilizing forecasts that give cognizance to risks and uncertainties involved.
NEED	Important for effective management, particularly in highly integrated networks, such as telephone, rail transportation and power transmission systems.	Crucial in all firms that feel the impact of sharpened competition for markets and the scientific-technological "take-off" with its ever-widening thrust.
FLEXIBILITY	Desirable, but accomplished primarily by relying on "patching up" from time to time. Systems were thereby adjusted to accommodate changes.	Highly desirable. System may not be complete unless it has been "scheduled" to allow for and even promote new inventions in products, processes and distribution.
DEMANDS	Managers and executives required to be thoroughly familiar with the methods, procedures and goals of systems in use in the organization.	Thorough knowledge required not only of systems in use but also systems methodology and technology, in order to work with systems engineers in the design of new systems or modification of old ones.
DECISION-MAKING	Incidental to but not within the scope of systems analysis.	Based on systematic analysis of alternatives and their interacting potentials and limitations. To the extent that net economic effects of each choice cannot be fully analyzed, a high degree of managerial judgement will be called for.

EFFECTIVENESS	Built in an element-at-a-time fashion, the complex of procedures, sub-systems and equipment lacks over-all plans and effective controls. Inconsistent elements may work at cross-purposes. Ultimate payoff is meagre.	Organized design, formally structured and disciplined, co-ordinating all man-machine and information-flows factors. Effectiveness is built-in. Total payoff may exceed the sum of the component contributions. Results and payoff may be highly rewarding.
EVALUATION	System value and effectiveness interpreted in terms of traditional cost accounting principles.	True economic value of system is evaluated in relation to various alternatives and ultimate effectiveness attainable, using advanced cost accounting and operations research techniques.
CONTROL	Primarily through review of accounting and cost records.	Control features imbedded in the original design. Continuous surveillance of actual versus aimed-at progress, so that goals are attained "on time" and "fully" to the extent possible.
INFORMATION FLOWS	May be haphazard, leaving dangerous gaps or entailing costly redundancies.	Integrated into the complex from the outset with checks to avoid missing links or needless replication.
INTEGRATED APPROACH	Primitive wherever attention to technological, human-factors and information-pattern aspects is compartmentalized.	Conscious aim-directed effort to equate engineering psychology, mathematical-statistical analysis and computer simulations in relation to the interlocking functions, feedbacks and controls pertinent to the system.
NEW EXPENDITURES	Justified as though each fragment were self-contained. If analyses are narrow, shallow and unstructured worthwhile projects may be viewed in limited scope and thus erroneously scrapped.	Justified on the basis of net economic value to the organization as a whole, based on a complete appraisal of the entire man-machines-informations complex. An objective determination of merits can thus be obtained.
ENDURANCE	May tend to downgrade unless patched up often. Where system obsolescence is rapid, the organization may incur the costs of antiquated methods and procedures.	Workable and lasting if breadth and problem-penetration during design were adequate. Periodic modifications and amplifications call for high managerial and technological competence.

FIG. 11-1 Modern systems methods and methodologies, and how they link together the business requirements of modern management with product, process and project engineering needs.

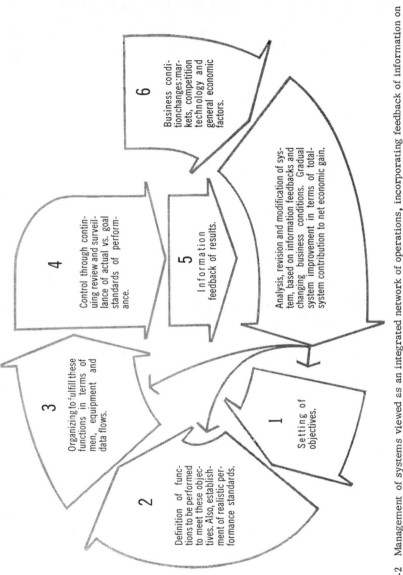

1 Setting of objectives.

2 Definition of functions to be performed to meet these objectives. Also, establishment of realistic performance standards.

3 Organizing to fulfill these functions in terms of men, equipment and data flows.

4 Control through continuing review and surveillance of actual vs. goal standards of performance.

5 Information feedback of results.

6 Business condition changes: markets, competition technology and general economic factors.

Analysis, revision and modification of system, based on information feedbacks and changing business conditions. Gradual system improvement in terms of total-system contribution to net economic gain.

FIG. 11-2 Management of systems viewed as an integrated network of operations, incorporating feedback of information on systems operation and performance as well as on general business conditions in their effect on the individual firm (see Fig. 11-3).

Mathematical programming for sales-production coordination fits well into these general systems integration concepts, as illustrated in Fig. 11-3.

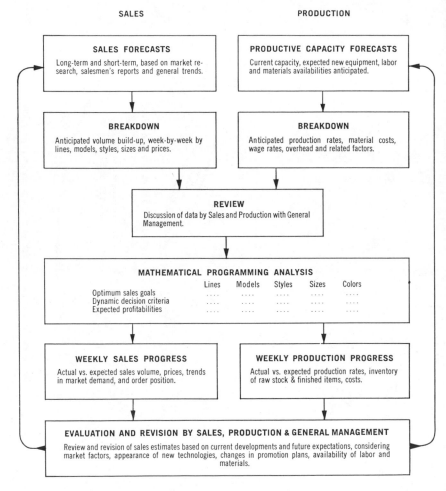

FIG. 11-3 Systems view of sales-production coordination aided by management science techniques: statistical and economic analysis of market trends and sales anticipation, inventory control, mathematical programming, and general quantitative analysis of interrelated factors.

Cost Evaluations

Comparative systems may be evaluated in terms of net economic gain, also known as "cost effectiveness," i.e., results attained vs costs expended. In such evaluations, certain principles seem to have gained favor among systems oriented managers:

1. Prior to a new decision, there will exist certain "sunk costs." If such equipment or facilities can be sold, then the net returns expected after dismantling may be applied toward a reduction of the cost of new investment. Otherwise, sunk costs are best disallowed.
2. It seems difficult to plan beyond 10 years. Risk and uncertainty factors rapidly increase with the length of anticipated futurity, and expected salvage value of equipment becomes harder and harder to assess.
3. In lieu of expected sales price, the worth of equipment in some alternative use is a very permissible substitute figure. The traditional accounting methods of amortization are usually quite inadequate for systems analysis.
4. Future costs and returns may be discounted based on anticipated net cash inflow. Usually, a low interest rate of 5 to 7% based on commercial market or long-term bond rates, results in an underestimation. On the other hand, utilization of 15 to 20% values, based on rate of return on corporate capital investment, seems excessive. Instead, a more suitable discounting rate would be based on a realistic estimate of the risk-free before-taxes return rate. This rate is usually somewhere halfway between the two sets of rate just cited.

In view of the enormous weight of business experience and judgment of managers, one usually encounters little difficulty in ranking relative cost choices. But neither the executive nor the systems analyst may be able to make precise quantifications of system effectiveness. As a result, the search for complete quantitative measures and analyses may become elusive at least in part, and systems design thus remains as much an art as it has become a science. This fact emphasizes the essential function of the manager. He must not only be intimately familiar with the substantive matter of his executive functions, but he must also know systems analysis concepts and techniques well enough to guide the systems analyst, to properly interpret his findings and to effectively apply the results.

Summary

Planning through MP entails features of forward anticipation and retrospective adaptation akin to problems of systems design and operation. The nature of MP and its inclusion in a total system have been reviewed.

12

SYSTEMATIC SEQUENCING
OF SCHEDULES

Ascertaining the optimal goals of coordinated managerial operations, as attained with the aid of Mathematical Programming, represents a major step toward a well-meshed overall program, which takes cognizance of the interlocking requirements of various business and industrial functions. Additional Management Science techniques can be brought into play next, thereby enhancing the details of day-to-day and week-to-week planning, decision-making, and control. An important and indeed one of the earliest tools of value in this phase of scientific management is the sequencing control board, originated about the first decade of this century by Henry L. Gantt and also known as the Gantt chart.

In the coordination of production and sales, for example, this board permits the systematic sequencing of production orders through manufacturing, so as to meet the promise-dates given to customers while at the same time making the best feasible utilization of available productive machinery, equipment, production lines, and operating personnel.

Although the principal example in the following will be with regard to the scheduling of production operations, it should be apparent that any program that calls for the coordinated planning of various phases of operation in an overall project, program, or activity can be sequenced in a similar manner.

Sequencing Control Board

Key to systematic sequencing is the Sequencing Control Board, shown in Fig. 12-1, in abbreviated form. (An actual board may contain from 50 to 150 machines, machine-groups,

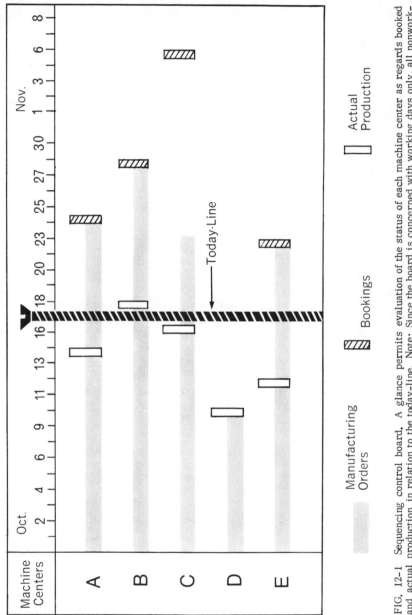

FIG. 12-1   Sequencing control board. A glance permits evaluation of the status of each machine center as regards booked and actual production in relation to the today-line. Note: Since the board is concerned with working days only, all nonworking days (Sundays, holidays, vacations, etc.) should be omitted from the time scale along the top.

or lines.) The board facilitates effective scheduling of production on various pieces of equipment on an hour-to-hour, day-to-day, and week-to-week basis. Quick adjustments in schedules can be readily made to accommodate customer changes, rush orders, and other contingencies or emergencies that may arise. The principle of the board can be extended to serve in a variety of applications, not only in production, but also in planning major steps in a research and development project, outlining the principal phases in a promotional campaign or scheduling the dispatch of service men in field maintenance problems.

As used in production, the Sequencing Control Board performs these principal functions:

1. Planning of work for the various machines, lines, or pieces of equipment. Figure 12-1 shows five Machine Centers, each Center representing a principal machine plus a number of minor auxiliary pieces of equipment. Planning is based on the production rate of the principal machine.
2. Knowing, by means of a glance at the board, the current status of bookings, manufacturing orders, and actual production in relation to the "today" line. Also, comparing bookings on hand with total productive capacity.
3. Estimating the dates at which deliveries can be promised for new orders. Again, a glance at the board will reveal the extent to which production on each machine is booked, and what space may be on hand to fit in additional orders.
4. Revising schedules to accommodate rush orders, when this can be done without violating existing commitments.
5. Noting, at a glance, any lagging production so that corrective follow-up can be initiated immediately.

The visual overview afforded by the Sequencing Control Board secures the foregoing objectives by means of a simple inspection of the status of orders and production.

Information Revealed by Board

The types of information presented at a glance by the Sequencing Control Board are illustrated for the various machine centers, as discussed below:

Machine Center A. Bookings in terms of customer orders extend until October 24. Production orders covering these bookings have been issued. Actual production, however, has been lagging. Today, on October 17 (as shown by the "today" line), production has resulted in an output that is the equivalent of manufacturing only to October 14. A check may reveal the causes, such as possibly maintenance or quality problems or other factors that are delaying the rate of output by comparison to standards.

Machine Center B. Actual production is slightly ahead of expectation. No problems.

Machine Center C. Actual production is quite close to expectation. Not all bookings have been translated into manufacturing orders. The reason for this will become apparent from a view of Machine Center D.

Machine Center D. There have been no bookings for some time, but a supply of stock was produced to inventory nevertheless. This stopped on October 10. Presently there is no production, because the equipment is being converted to make it suitable to produce some of the orders now booked for Machine Center C (which has a slight overload and is expected to be oversold if the market continues strong).

Machine Center E. Actual production is about eight days behind standard. Despite management follow-up, problems in meeting schedules have been plaguing the principal machine in this center. Customers were notified of unavoidable shipping delays.

Although the information revealed by this board could have been provided from the records maintained in most production planning and control departments, the visual at-a-glance data of the board gives quicker information and also readily reflect various interrelationships among all of the machinery and orders.

Machine Loading Cards

Each of the shaded row segments representing manufacturing orders, on the Board, consists of a number of individual Loading Cards, as illustrated in Fig. 12-2. These cards are inserted in transparent pocket holders of the board, with one pocket overlapping the other, but leaving the lower (shaded) portion visible. By lifting the pockets on the board, the full detail of each loading card comes into immediate view.

Machine Loading Card		
Shifts per day: 1 2 3 (check one) O O O	Machine center:	
Manufacturing order no.	No. of units to be made:	
Production rate:	Shifts per unit:	Days per unit
Date started:	Date due:	Time lost:

Time scale, cut to proper length

```
  2     4     6     8     10    12    14    16    18    20    22    24    26
 |  |  |  |  |  |  |  |  |  |  |  |  |  |  |  |  |  |  |  |  |  |  |  |  |  |
```

FIG. 12-2 Machine loading card. When completed, this card is inserted in the transparent pocket holders of the sequencing control board and folded under successive rows, so that only the lower portion containing the time scale is visible.

Shifts Per Day. Statement whether the particular machine center will be on one, two, or three-shift operation.

Manufacturing Order. The number of the manufacturing order covering this item. If desired, the Machine Loading Card may also provide space to indicate the customer orders going into one manufacturing order.

Production Data. The card not only indicates the shifts per day that a machine is operated, but also the number of units to be produced and the standard production rate. If, for example, a machine is on three-shift operation with a production rate of six shifts per unit, then this represents $6/3 = 2$ days per unit. Furthermore, at two days per unit, if 20 units are to be made, we will require a total of 20/2 or 10 days. Accordingly, the time scale along the bottom section of the card would be cut to show 10 days. It is essential, of course, that the length of each day on this scale correspond to the length given in the calendar scale of the Sequencing Control Board.

Dates. The date started, the date production is due to be completed, and a record of any time losses may also be shown.

The purpose of the loading card now becomes clear: it provides useful and often essential detail jointly with the Sequencing Control Board. In fact, by inserting each card in its proper place in the visible section of the board pockets, an integrated information source is provided. Changes in scheduling, shifts in assignment of orders, or rearranging of jobs among machines all can be expedited by simply moving the cards among the various pockets of the board. The result will be a well-planned sequencing of production orders which makes optimal use of available capacity while yet fulfilling promised delivery dates for completed units.

Nomograph

A convenient means of cutting the time-scale on the loading card to proper length, without any calculations, is provided by the special nomograph in Fig. 12-3. For our previous example involving the production of 20 units at two days per unit, we would proceed as follows:

1. Enter the Nomograph at the level of 2.0 for "Production rate, days per unit produced."
2. Place the loading card flush with this point, with the time scale just above the 2.0 line of the nomograph.
3. The diagonal line 20, for "Number of units to be produced" will cross the time-scale on the loading card at point 10.
4. Mark this 10 as the length of the time scale to be used. Cut off the remainder of the card scale.

With these steps accomplished, the loading card will now be of proper length for insertion in the Sequencing Control Board.

Project Sequencing

Networks comprising entire projects—such as the building of a new processing department, warehouse or plant, or the development of a promotional program, or the engineering of a new, complex equipment—benefit from the type of scheduling chart presented for production departments. Figure 12-4, for example, shows the network of activities planned for the construction of a specially designed electrical equipment. While the flow of activities is self-explanatory, the following comments are of interest:

1. The project flow chart indicates clearly what activities must precede subsequent work. For example, before overall assembly (G) can proceed, the chasis must have

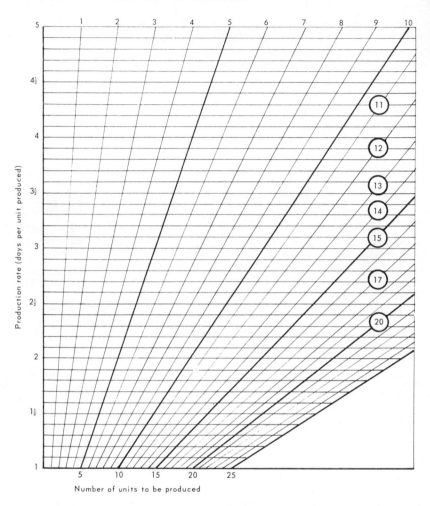

FIG. 12-3 Nomograph for determining length of time scale on machine loading card.

been constructed, (F), the power supply built (E), and the components assembled (D).

2. Unused, idle, wait, or "slack" time may occur. For example, Activities B and C must precede D. B takes four days but C uses only two days. The slack is thus $4 - 2 = 2$ days. Slack time represents opportunity for improvement. For example, if our overall planning of manpower assignments, facilities usage, or financial arrangements should make it desirable, we could move Activity C from days 6 and 7 to days 4 and 5 or days 5 and 6. Slack time is shown in dotted lines.

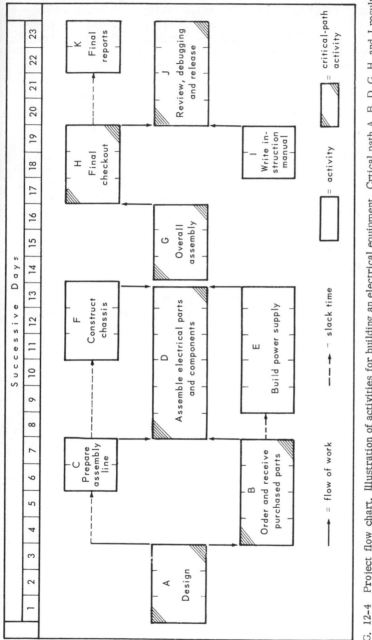

FIG. 12-4 Project flow chart. Illustration of activities for building an electrical equipment. Critical path A, B, D, G, H, and J results in a time requirement of 3 + 4 + 6 + 3 + 4 = 23 days expected from start to finish of project. Slack time represents flexible scheduling opportunities. For example, it may be possible to move some critical-path work into the slack, if this reduces overall time requirements. Also, slack time may permit some moving of activities within this allowed time, which may result in better use of manpower, facilities or financial resources. Reviews and revisions of these networks will result in successive improvements toward a highly efficient final schedule in terms of time and cost factors. (Other terms for this type of chart are PERT for Program Evaluation and Review Technique or CPM for Critical Path Method.)

3. The path traced by a sequence without any slack lines is the *critical path*. In our example it signifies that we cannot under present plans expect less than 23 days to completion. The critical path is thus the slackless trace through the diagram.

4. Often, the time involved in the critical path can be reduced. For example, it might be possible to begin some of the assembly of components in Activity D from available parts, while we are waiting for purchased parts (Activity B). In turn, this calls for moving D back to start on day 6 rather than day 8, while C moves likewise by the two slack days. Other activities, such as E and F and all other activities would similarly move to earlier days. Total time requirements are thus reduced.

5. Another way to reduce overall time is to examine the critical path closely for places where the use of additional manpower or equipment might produce a significant shortening of time. For example, Activity D might be cut in half to only three days by doubling the assembly operators and facilities. Obviously, this will remove D from the critical path, while E takes its place. So, thought must next be given to reducing time for E.

6. At all times, the factors of possible shortening of project length must be weighed in terms of (a) urgency of need for the completed item, be it a piece of equipment or a new plant and (b) the cost involved in speed-ups. The evaluations are complex, because lessened time to complete projects may yield savings in terms of reduced interest charges on borrowed money until the project is operational and profitable, and these savings may tend to exceed the extra costs of overtime, rush-schedules, and special assistance. The network provided by the project flow chart is invaluable in these evaluations.

Project Cost Flow

A slight modification of our activity network will reveal the flow of cost. For simplicity of illustration, only the project labor costs appear in Fig. 12-5, but it is apparent that materials and other items involving financial outlay could be easily included. The procedure is readily apparent:

1. Break each activity into its daily cost. For example, if Activity A involves a labor outlay of $300 in three days, then the daily cost is $100. The total cost for days 1, 2, and 3 each is $100, since A represents the only activity.

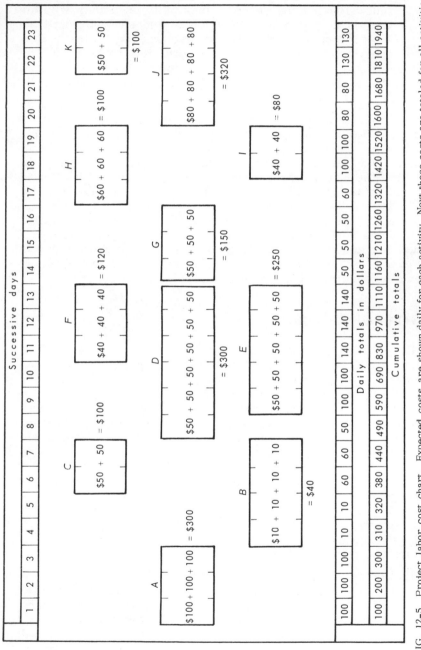

FIG. 12-5 Project labor cost chart. Expected costs are shown daily for each activity. Next these costs are totaled for all activities per day and then cumulated. Actual cost accumulations can thus be checked against this schedule of expected costs.

2. Cumulate the daily totals. Thus, $100 for the first day plus $100 for the second gives $200, a further $100 for the third day cumulates to 300, and so forth, until the final project cost of $1940 is reached.
3. Daily totals may consist of several activity entries. For example, on day 11, costs are incurred by activities F, D, and E of $40, 50, and 50, totaling $140 for the day.

The cumulative values are especially valuable. They represent currently expected and thus scheduled costs. If actual costs exceed schedules, a check is warranted. Why are they excessive? The reason need not be poor cost performance. Often a project proceeds at a faster pace than originally envisioned. So long as the accelerated progress is consistent with allowed expenditures, no cost problem exists. For example, assume that after 10 days of actual work the progress made is equivalent to 12 days originally scheduled. The scheduled cumulative costs were $690 for 10 days and 970 for 12 days. Therefore, if actual accumulated expenditures are at or below $970, there is no cause for concern. On the contrary, the project manager should be congratulated for being ahead of schedule.

Psychological Value

Beyond the fact that plans, schedules, programs, and visual boards or charts contribute to a more orderly, efficient, and knowledgeable manner of doing business, we should not overlook the direct motivational value involved.

Managers, foremen, supervisors, and operating personnel have been given a definite schedule, a goal standard. If this set of time and cost values has been developed with their assistance and active cooperation, then it is likely to be a fair one and to be recognized as such. Next, it becomes great fun to try to beat the schedule. Often this can be done more with thoughtfulness and ingenuity than brute hard work. For example, in the development of a complex equipment, the project manager and his staff proudly completed the job some 20% ahead of schedule with an expenditure well below original anticipations. Several factors contributed, all attributable to effective teamwork:

1. Design engineers working with production people were able to come up with revisions that enhanced quality while reducing assembly time. Although the revisions were minor, their effect was significant in terms of costs.
2. The production engineer in charge found that he could assign to well-qualified and reliable technicians many of

the tasks that had originally been assigned to be handled by engineers. The engineers, in turn, were thus available for other urgent projects.

Finally, as experience was gained with estimating time, materials, staff requirements, and labor needs, it was possible to develop improved project flow charts. Not only in terms of morale within the organization, but also as regards customer response to the firm's consistent ability to meet deadlines on critically needed special equipment, these psychological gains were immeasurable and ultimately found their reflection in the overall profitability and growth data.

Summary

The sequencing control systems and visual aids, such as depicted in this chapter, provide a valuable adjunct to managerial programming. They permit detailed scheduling of phases of activity with an effective visual overview. Improved planning and performance results. In those instances where for any reason actual progress lags behind plans, the control boards and charts spotlight such occurrences for early check and investigation. Timely corrective action will thus be possible.

13

MINIMIZING THE EFFECT
OF UNCERTAINTY IN PLANNING

Providing for the future involves many uncertainties regarding economic conditions, market demand, manpower availability, and materials costs. What prices will we be able to charge? What will be the market for these new products? Can we get the materials and manpower needed? What will the competition be coming out with? All these are weighty questions involving a good amount of uncertainty. Obviously we cannot foresee the future, but we can try to develop predictions to the best of our abilities. Fortunately methods have been developed whereby the planner can effectively cope with the factors of uncertainty, thus minimizing their possible adverse effect on the value of programming.

Risk and Uncertainty

To understand the nature of uncertainty, we may first examine a closely related term, "risk." For example, if experience has shown that flying a certain type of small airplane in a snowstorm has in the past resulted in one crash per thousand flights, then the risk involved in the next flight for you—assuming you use a randomly chosen or just any plane, and the pilot is not really different in skill from the average—is 1/1000 or 0.001. Risk is a familiar term in personal and business life, and insurance rates are generally based on experience rates of the nature just discussed.

In many situations involving risk, no nice and neat experience rates are available. Will a new product catch on? Is the planned venture likely to be successful? How close is a research project toward payoff? What will the future economic and market conditions be like? Answers to these questions involve evaluations in terms of managerial experience, the executives' considered judgments or opinions. Objective data, such as trend projections, correlations with anticipated developments in related areas, or consumer research, all will aid in this judgment. But the plain, simple rate of the insurable type of risk—fire, theft, windstorm, crash, or bad debts—all have vanished. We speak of this type of noninsurable risk as "uncertainty."

For a further distinction, we may look upon the simple type of risk as "objective," in the sense that it involves merely the rate taken from adequate past experience, which is likely to prevail in the near future. The probability of an airplane crash, under the conditions discussed earlier, is 0.001 or one-tenth of 1%. It is an objective probability. On the other hand, if an executive says: "I have discussed this research project with the director of research, and he thinks there is an 80% chance that it will be completed successfully in the next six months" he is talking about a subjective probability. It is subjective because it is based on judgment—personal opinion, supported by experience in similar types of problems or situations, and an evaluation of all pertinent factors by the manager, executive, or staff man. The joint judgment, obtained by consensus or averaging of percentage estimates of individuals comprising a group, is also a subjective probability.

The executives, managers, or staffers may sharpen their judgment by relying on statistical extrapolations of trends, correlations of related data series, consumer surveys, and other objective data. So long as the principal evaluation is judgmental, the probability given—such as the likelihood of success of a new venture or the chances that a research project will pay off—is to be considered subjective and the type of chances taken are considered in the realm of "uncertainty" rather than risk.

Admittedly situations may arise where borderline questions occur and where one may argue whether we are dealing with risk or uncertainty. But an unhealthy preoccupation with definitions would not be called for. So long as we know the distinctions, we need not worry about the instances of convergence of risk and uncertainty. A more fundamental question that may be raised is whether it is valid to treat a judgmental opinion, such as, "there is a 90% probability that we will be able to sell next month's output without lowering prices," as a probability at all. Probability, in the classic sense, depends on verifiable past outcomes—such as the throws of unbiased dice or the life expectancy rates of humans. An opinion is based on experience in situations similar to the one at hand at a particular time. But the judgments and weights and past data exist primarily in the opinion-maker's mind and are hardly verifiable in an objective way. Despite these differences between objective and subjective probabilities, and the obvious superiority in precision of the former over the latter, there seems no alternative to using subjective probability* when risk-factors and objective values are not available. The words "not available" are used advisedly. Instances may arise when, theoretically, objective data could be gathered. In practice, however, this process may involve undue delays while a decision must be made now based on data presently obtainable.

Objective Probability

As an example of objective probability, we will examine a classic case of the toss of two unbiased coins. The various possible outcomes can be diagrammed in the form of a "probability tree" in Fig. 13-1. Attaching probabilities for heads and tails of 50% or 0.5 in decimal form to each individual toss, successive multiplication of probabilities gives the familiar expectation of:

> Both coins come out heads, 25% or 0.25.
> Both coins come out tails, 25% or 0.25.
> Head and tail combinations, 50% or 0.50.

We may consider these percentages as the expected values for the various possible outcomes. It is axiomatic that they must total to 100% if we have considered all possible outcomes.

Subjective Probability

Instead of a coin, let us now look at a business venture. We are out to develop a new synthetic material and feel that a certain general avenue of approach has an 80% chance of success. Only one other firm has the resources to do similar research, but while they seem to prefer another avenue of approach (based on their judgment), we feel subjectively that they have only a 70% chance of success. Needless to say, our success and failure probabilities are based on the experience judgment of the chief research supervisor and his senior staff, supplemented by literature searches and certain pilot studies.

Although we now deal with subjective probabilities, we may use them in a manner quite parallel to the coin-tossing procedures, as in Fig. 13-2. The results give the various expected

---

*If this were not so, there would never be any "betting odds" on horse races or prize fights.

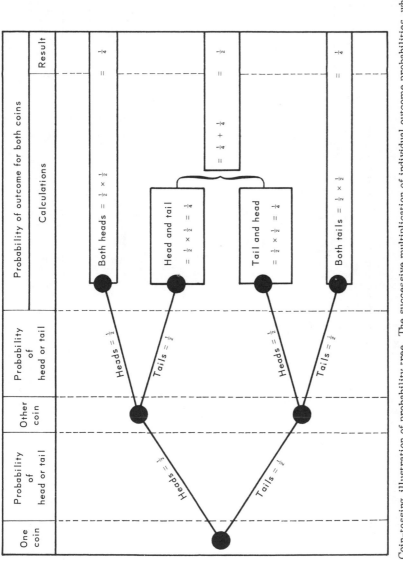

FIG. 13-1   Coin-tossing illustration of probability tree. The successive multiplication of individual outcome probabilities, when put in diagram form, resembles a tree. Hence the term "tree diagram" or "probability tree." For the two coins we find these probabilities: Both heads = 1/2 × 1/2 = 1/4, similarly for both tails, and for heads-tails combination = (1/2 × 1/2) + (1/2 × 1/2) = 1/4 + 1/4 = 1/2. Extensions to three or more coins are readily apparent.

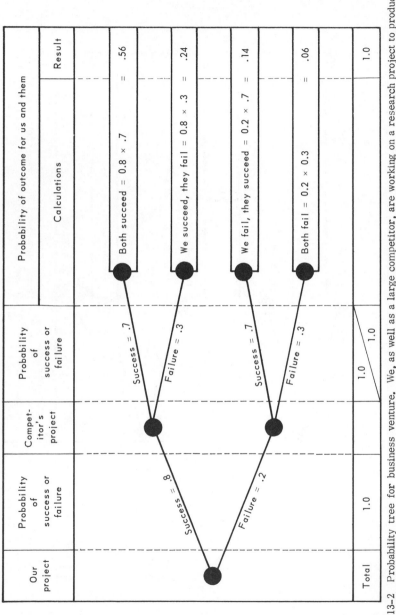

FIG. 13-2 Probability tree for business venture. We, as well as a large competitor, are working on a research project to produce a new synthetic material; we each utilize a different approach. Success and failure probabilities, based on available competent opinions, are given. The resultant outcomes are then obtained through multiplication of individual probabilities. For example, there is a 24% chance that we will succeed while the competitor will not (second bar, above).

outcomes, which may now be invested with cost and profit calculations as an aid in decision making.

Cost and Profit Calculations

The following are the basic data involved. The cost of the research is estimated at $4 million. Successful development of the new material should be worth $10 million in terms of sales profit. This, as well as all other cost data, have been "discounted to present value"; which means that, even though the costs and sales revenues will occur over varying periods of time into the future, their ultimate value can be determined for the present in terms of a generally acceptable interest rate. For example, if cash is worth 10% per year, then $1.10 to be received 12 months from now is worth only $1.00 received today.

We have also made an agreement with our competitor. If he fails to succeed in his research, he will buy a license to use our process for two million dollars. Conversely, if we fail but he succeeds, we will license from him for the same amount. If we both succeed or both fail, no licensing fees are paid to anyone. Here are some of the basic calculations for the anticipated dollar profits (in millions) for the various outcomes:

1. We and he both succeed. Sales profit of 10 less research of 4, net 6.
2. We succeed, he fails. The 6 is increased by 2 because of the licensing fee paid us.
3. We fail, he succeeds. The aforementioned $6 million is now reduced by the $2 million licensing fee to $4 million.
4. We both fail. There is no product to sell, but the 4 of research costs has been expended and thus represents a loss.

These costs are entered in the middle section of Fig. 13-3, with the basic probabilities for the various outcomes, derived from the probability tree diagram, in the first section.

Expected Value

We now need to determine the expected values associated with this research venture. Before we proceed with this, let us demonstrate this concept from a simple example.

		1	2	3	4	5	6	7
		COMPETITOR						
		PROBABILITY (from tree diagram)		Anticipated profit to ourselves, in $ million		Expected profit to ourselves, in $ million		Expected value, $ million
		Success	Failure	Success	Failure	Success	Failure	
				*	*	(1) × (3)	(2) × (4)	(5) + (6)
a	Ourselves — Success	0.8 × 0.7 = 0.56	0.8 × 0.3 = 0.24	6	8	0.56 × 6 = 3.36	0.24 × 8 = 1.92	5.28
b	Ourselves — Failure	0.2 × 0.7 = 0.14	0.2 × 0.3 = 0.06	4	−4 (loss)	0.14 × 4 = 0.56	0.06 × (−4) = −0.24	0.32
c	Total expected value to ourselves							5.60

FIG. 13-3 Expected value of business venture. A research project is undertaken by ourselves as well as the competitors. The latter utilize a different approach. By agreement, if one succeeds and the other fails, the one failing may purchase a license at $2 million. Other anticipated costs and profits are as shown. Adjustment of these values to allow for probabilities of success or failure results in the expected values shown. The total expected value of the venture (to ourselves) is $5.60 million, but as shown a maximum profit of $8 million or a loss of $4 million are possible. The expected value is the most likely profit.

*Based on the following data, in millions of dollars, discounted to present value terms: Estimated research cost, 4; estimated sales profit, 10; licensing fee, 2. The latter is paid to us, if we succeed but the competitor does not; or paid by us in case he succeeds and we fail. If both succeed or both fail, no licensing fee is involved, hence the net loss of 4.

A salesman is considering a trip, involving a $100 cost, to visit a customer who might buy a heat pump from him at a gross profit of $500. Past experience, however, shows that of 10 sales presentations only one has lead to a sale. The probability of a sale is thus 10%, and the expected gross profit is 10% × $500 or $50. After deducting the $100 cost, the net profit or value of the trip is a loss of $50. Clearly, in the absence of data other than those given above, the trip should not be made. The concept of expected value leads to this decision.

The expected profit for our research project can be evaluated similarly, as shown by the calculations in the last three columns of our tabulation. An expected value of $5,600,000 results. Admittedly if we are very lucky—that is, we are successful but our competitor is not—we will make $8 million. We also note a $4 million loss in the extreme bad luck case. The expected value means that, in the long run, if we continue to use a system of evaluation such as presented, we will tend to obtain the type of results reflected by the total expected value.

There is a need for careful exercise of judgment in developing the initial subjective probabilities. Moreover, to the extent possible, objective probabilities should be used. Also a problem may contain many more factors and facets than in the example given. But the principle involved remains valid: A system has been developed that permits logical use of the executives' judgment in evaluating the expected values of a project, venture, or other decision-requiring activity.

Expected Value—Further Example

The possible decisions that can be made or that we wish to consider may be viewed as strategies. For example, a mail order house in printing its catalog may have to consider a variety of markups for certain product categories. Whether or not these markups will be optimal will depend, among other things, on volume attainable. In turn, price, volume, and type of season may interact, thus resulting in different profit anticipations (first two sections of Fig. 13-4). Assume now that a combination of economic research and opinion polls among the responsible executives and staff men has resulted in the subjective likelihoods of 20, 60, and 20% for a good, fair, or poor season, respectively (lower section of the figure). Then, multiplication of the anticipated profits by the probabilities stated yields the expected profits shown. Upon summing, these expected values result:

Markup, %	Expected value, $
10	5,400
20	5,000
30	6,000

The strategy of 30% markup, resulting in the highest expected value, would be chosen. However, there are other strategy approaches. A firm that is in financial difficulty,

	(1)	(2)	(3)	(4)	(5)
	STRATEGIES	STATES OF NATURE			
		Good season	Fair season	Poor season	
	Markup, %	Anticipated sales, $1000 per week			
a	10	100	50	20	
b	20	35	25	15	
c	30	30	20	10	
		Anticipated profit, $1000 per week = markup % × anticipated sales			
d	10	0.10 × 100 = 10	0.10 × 50 = 5	0.10 × 20 = 2	
e	20	0.20 × 35 = 7	0.20 × 25 = 5	0.20 × 15 = 3	
f	30	0.30 × 30 = 9	0.30 × 20 = 6	0.30 × 10 = 3	
		Probability, %, of good, fair or poor season			Expected value of strategy, $1000 per week
		20	60	20	
		Expected profit, $1000 per week = probability % × anticipated profit			= (2) + (3) + (4)
g	10	0.2 × 10 = 2.0	0.6 × 5 = 3.0	0.2 × 2 = 0.4	2 + 3 + .4 = 5.4
h	20	0.2 × 7 = 1.4	0.6 × 5 = 3.0	0.2 × 3 = 0.6	1.4 + 3 + .6 = 5.0
i	30	0.2 × 9 = 1.8	0.6 × 6 = 3.6	0.2 × 3 = 0.6	1.8 + 3.6 + 0.6 = 6.0

FIG. 13-4 Expected value of three markup strategies. Based on the data given, the 30% markup yields the highest expected value in terms of profit per week, viz., $6,000.

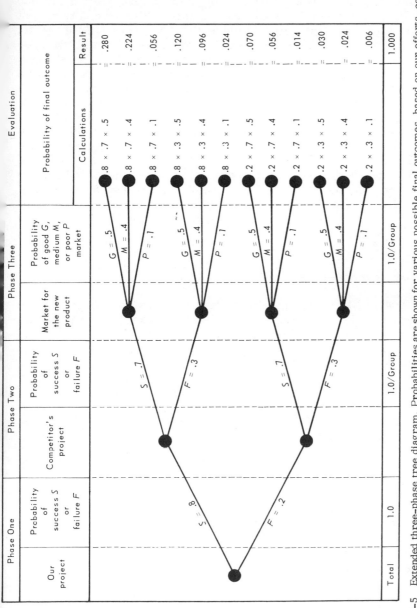

FIG. 13-5   Extended three-phase tree diagram. Probabilities are shown for various possible final outcomes, based on our efforts, competitor's efforts and market conditions expected to be encountered for new product.

likely to be wiped out unless the next season is highly profitable, would "go for broke." The 10% markup can yield a $10,000 profit, if the firm is lucky enough, viz., if the season turns out to be a good one, despite the mere 20% probability of this state. But this strategy would hardly suit another type of timid management bent on providing against the worst case. The worst case, of course, is that state representing a poor season, where a 10% markup results in the lowest profit of $2,000. Strategies of 20 and 30% markup, each giving $3,000 under a poor season, are then preferable. We have labeled the seasons "states of nature" because there is nothing that management can do to bring about their appearance. Like the weather, it is out of management's control.

Market and other pertinent factors can be added to any decision-making problem. For example, the research venture tree diagram previously discussed could have been enlarged to consider the type of season along with competitive actions, as in Fig. 13–5.

Using Uncertainties in Programming

Probabilities relating to uncertainties can be readily introduced into the programming problems of management planning. Let us examine the data for a firm's cold rolled steel production and marketing in Table 13-1. For Light Gauge Steel, for example, the Sales Department is practically certain (that is, virtual 100% probability) that up to 600 rolls can be sold per day. There is furthermore a 50% chance of the sale of an additional 200 rolls, making a total of 800 per day, and a 20% chance of getting up to 900. Lines a and b represent these evaluations. The increments of 200 and 100 rolls, above the basic 600, as just discussed, are next stated explictly in line c. Based on anticipated attainable prices and profits per roll, we are able to compute the expected profits at various increments of market demand, as in line f.

Finally, using the current production rates in the two principal processing departments, rolling and annealing, and their respective capacities in Table 13-2, we arrive at the standard form of a Mathematical Programming problem. The optimum program is thus readily obtainable.

Summary

Uncertainty in planning requires judicious application of subjective probabilities. Although based on personal judgment,

Table 13-1 Distribution of Expected Market Demands and Values
for the Firm's Products

Initial data and calculation steps*	Column 1	Column 2	Column 4
a. Possible market demands, rolls per day			
Light gauge steel, L	600	800	900
Heavy gauge steel, H	300	400	450
b. Probability, %, that market demand above will be attained	100†	50	20
c. Incremental rolls per day‡			
Light gauge steel, L	600−0 =600	800−600 =200	900−800 =100
Heavy gauge steel, H	300−0 =300	400−300 =100	450−400 =50
d. Expected markey demand (= b × c)			
Light gauge steel, L	600	100	20
Heavy gauge steel, H	300	50	10
e. Profit per roll, $			
Light gauge steel, L	10	10	10
Heavy gauge steel, H	20	20	20
f. Expected profit per roll at various increments (= e × b), $			
Light gauge steel, L	10	5	2
Heavy gauge steel, H	20	10	4

* Lines a and b based on sales department estimates. Line e derived from sales department estimates and cost accounting data, utilizing expected prices and costs.

† Practical certainty is shown as 100%.

‡ Increments are obtained from line a. Column 1 is the same in both lines a and c. For all others, the nearest left-hand column entry is subtracted from the given column entry. For example, column 2 entry of 800 minus 600 of the prior column 1 gives an incremental market demand of 200; viz., the increment of col. 2 over 1.

these values can nevertheless be logically and usefully interpreted within the framework of mathematical probability and MP analysis. If the basic judgment is good, based on sound evaluation of subjective and objective factors, then the programming will yield data that serve in decision-making that, in the long run, is bound to be superior to any other approach.

Table 13-2 Programming Problem Involving Uncertainties Regarding Market Demand

Problem data	(1)	(2)	(3)	(4)	(5)	(6)	(7)
	Saleable steels						Available capacity hr/day
	Light gauge, L			Heavy gauge, H			
	L-1	L-2	L-3	H-1	H-2	H-3	
a. Expected profit, $/roll	10	5	2	20	10	4	
b. Production rate, hours per roll							
Rolling Department	0.8	0.8	0.8	0.4	0.4	0.4	400
Annealing Department	0.2	0.2	0.2	0.4	0.4	0.4	200
c. Expected market limitations, rolls salable per day	600	100	20	300	50	10	
d. Optimal program, no. of rolls	333.3	...	...	300	33.3	...	

Source: Lines *a* and *c* from Table 13.1. Line *b* from Standards Department data.

Notes: From a customer's viewpoint, only two steels, *L* and *H*, are sold. However, because of uncertainty (decreasing probability) of selling beyond the 600 and 300 rolls of "practical certainty," (line *c*), we have for analysis purposes also products *L-2, L-3, H-2*, and *H-3*, with profits and quantity limitations adjusted for these decreasing probabilities. Details of adjustment appear in the table referred to under "Source" above.

14

SOLVING LARGE–SCALE
INTERLOCKING PROBLEMS

As has already been indicated, for the solution of large-scale problems involving interdependent variables, we rely on a series of so-called matrix steps. The term "matrix" refers to the orderly arrangement of data in a block of figures or "tableau," examples of which will be given. The particular most universally applicable approach is known as the Simplex method, originated by Dantzig,* and this will be used here.

Although large-scale problems are handled most quickly and inexpensively by a computer, nevertheless many readers may be interested in the matrix algebra of MP.

Illustrative Example

For the purpose of demonstrating the matrix steps, we will utilize a previous example (Chap. 2) of a firm concerned with machining and assembly of only two products, A and B, with profits (contribution to profit and overhead) of $10 and $8 per unit, respectively. Capacities are 240 and 80 hr per week in machining and assembly, respectively. Product A requires 6 hr of machining and 4 hr of assembly per unit and product B requires 8 and 2 hr, respectively. A graphic solution was shown, and it will be interesting now to observe the matrix approach. Elementary algebra alone is needed.

Steps

We begin by expressing our goal or objective, to maximize profit, in the form of an equation. Next, the factors that limit

*G. B. Dantzig contributed chapters in T. C. Koopmans (ed.) Activity Analysis of Production and Allocation, Cowles Commission Monograph, no. 13, J. Wiley & Sons, New York, 1953.

our alternatives—production rates, available capacity—are similarly expressed. The procedures are given in detail in Table 14-1: "Profit-Maximizing Problem Expressed in Equation Form."

Next the equations are reshaped in a special tabular form, known as a matrix. From the first or initial matrix, a series of further matrix steps yields a final matrix that contains the optimum as the solution. Figure 14-1, together with the text provided with this material, shows the matrix transformations involved. The text is purposely held to a minimum. It is simpler to follow the sequence of operations—aided, if you wish, with pad and pencil—than to try to wade through lengthy explanatory writing.

Note that Figs. 14-2 and 14-3 provide supplementary illustrative explanation of the matrix steps.

The final matrix yields the solution, which is noted from the rows labeled B and A for products B and A. In the Result column, we are told that to maximize profit we should aim at 24 units of B and 8 units of A per week, which at the contributions to profit of $8 and $10, respectively, will yield (8×24) + (10×8) = $272 profit, as indicated in the Z row.

Systematic Search for the Optimum

The matrix steps provided a systematic search for the optimum. We began with zero profit (products M and N in the far left-hand column yielded a total profit of 0, as shown by result R for row Z).

In the initial matrix, the value of -10 for $Z-C$ of column A told us that we were losing the highest amount of profit opportunity by not making that product. (In the initial matrix, products M and N are made.) This led us to further investigations that said, in effect: "Substitute product A for product N as the next step."

The second matrix indicated that this substitution had increased profit to $200. But we also noted that we were losing profit opportunity by failing to make any product B. Substituting product B for product M brought us to the final matrix, with a maximum profit of $272.

This sequence of steps is further emphasized by the diagram of the results of this profit search, in Fig. 14-4. Certainly, the steps followed so tediously in mathematical fashion could have been traced relatively simply by ordinary trial-and-error. Let us not forget, however, that if a large number of products and processes are involved, any trial-and-error approach will only

Table 14-1 Profit-Maximizing Problem Expressed in Equation Form

1. Goal Equation

The goal is to maximize dollar profit, Z, based on the profit contributions of \$10 and \$8 per unit of products A and B produced. Therefore, we need to know that quantity A of product A and that quantity B of product B which will satisfy the goal or objective:

$$10A + 8B = Z_{max} \qquad \text{Eq. (1)}$$

2. Production Equations

In machining, production rates are 6 and 8 hr per unit for products A and B, respectively. In assembly, the corresponding rates are 4 and 2. Total production cannot exceed and thus must be *equal to or less than* (shown mathematically by the symbol $\leq$) the capacity of 240 and 80 hr in the two departments.

$$\text{In machining:} \quad 6A + 8B \leq 240 \qquad \text{Eq. (2)}$$

$$\text{In assembly:} \quad 4A + 2B \leq 80 \qquad \text{Eq. (3)}$$

3. Achieving Symmetry

The inequality signs $\leq$ are messy. But we can make equality signs out of them by adding suitable magnitudes to the left-hand side of (2) and (3) above. These magnitudes may be called "imaginary variables" of size 0 or greater. Using symbols M and N for these imaginaries and inserting zero coefficients as shown below, we obtain symmetrical equations for (1) to (3):

$$10A + 8B + 0M + 0N = Z_{max} \qquad \text{Eq. (1a)}$$

$$6A + 8B + 1M + 0N = 240 \qquad \text{Eq. (2a)}$$

$$4A + 2B + 0M + 1N = 80 \qquad \text{Eq. (3a)}$$

A value of $0M$ or $0N$ will, of course, be equal to zero. Nevertheless, we show these values here, in order to obtain equations that are symmetrical in that an A, B, M, and N appear in each equation, even though the value of M and N may happen to be zero.

4. Solving for the Optimum

The equations (1a) through (3a) are now cast into a matrix, as shown in the first Initial Matrix of Fig. 1. Note that we have raised the variables A, B, M, and N to the column tops and have placed under each column, row by row, the coefficients of (1a) through (3a). The matrix so formed represents a convenient method for solving for the maximum profit program.

lead to hopeless confusion. The matrix procedure, on the other hand, plods along unerringly until the maximum profit (or minimum cost) has been attained.

	Contri- bution, C →	Products				Result R	Evaluation R ÷ A	Key Row
		A	B	M	N			
		10	8	0	0			
INITIAL MATRIX (PROBLEM)	M	0	6	8	1	0	240	40	
	N	0	4	2	0	1	80	20	←
	Z		0	0	0	0	0		
	Z–C		–10	–8	0	0			
	Key Column	↑						

	Contri- bution, C →	Products				Result R	Evaluation, R ÷ B	Key Row
		A	B	M	N			
		10	8	0	0			
SECOND MATRIX	M	0	0	5	1	-1.5	120	24	←
	A	10	1	0.5	0	.25	20	40	
	Z		10	5	0	2.5	200		
	Z–C		0	-3	0	2.5			
	Key Column	↑						

	Contri- bution, C →	Products				Result R
		A	B	M	N	
		10	8	0	0	
FINAL MATRIX (SOLUTION)	B	8	0	1	.2	-.3	24
	A	10	1	0	-.1	.4	8
	Z		10	8	.6	.16	272
	Z–C		0	0	.6	.16	

FIG. 14-1 Matrix steps leading to maximum profit solution. Explanation for each of the three matrices is given in Table 14-2.

An Extended Model

In order to illustrate an extended application of the Simplex method, let us assume that a firm has products X and Y for

Table 14-2 Matrix Steps

Initial Matrix

Columns A to R show the numerical coefficients of Eqs. (1a) to (3a), for row C representing Contribution to Profit, to rows M and N representing the imaginary products.

Row Z entries for each column are obtained by multiplying the column entries by the Contribution to Profit, C. For column A, for example, $6 \times 0 + 4 \times 0 = 0$, which is entered in Z. Finally, row C is subtracted from Z.

That column which contains the lowest $Z - C$ is called the Key Column. Since $Z - C = -10$ for column A, A is the Key Column. The Key Row is the one that contains the smallest positive ratio R/A, which turns out to be N. The intersection of Key Column and Row is the Pivot, which is 4 in our example. It represents the lowest $Z - C$ and the lowest R/A, and thus forms the basing point from which a new matrix will be developed. This new matrix will seek to increase profits until a maximum has been found.

Second Matrix

Row M, columns B to R is found by transforming the entries of row M of the initial matrix, using the method in Figs. 14-2 and 14-3. Column A, containing the Pivot, becomes 0 for all Product Rows except that the pivot always becomes 1. The other entries in Row A are found from Row N of the initial matrix, by dividing each row value from columns B to R by the pivot 4. Under column C, the unit profit contribution of \$10 is shown, thus indicating that product A has been substituted for product N.

Rows Z, $Z - C$, and the new Pivot are found as before. Only when all of the entries in $Z - C$ are zero or positive will the optimum solution have been found. Although current profit is \$200, from row Z column R, we have not yet found the optimum, since $Z - C$ for product B is negative. Another matrix must be formed.

Final Matrix

Obtained by steps paralleling those that yielded the second matrix, the third matrix contains no negatives in the $Z - C$ row. We thus know that a maximum has been reached, which is \$272 as row Z, column R shows. Also, as column R indicates, weekly quantities of 24 units of product B and 8 units of product A will lead to this maximum. Since this third matrix has yielded our desired solution, it is the final matrix.

---

sale at profits of \$2 and \$5 per unit, that available capacity is 45 hr per week and that the production rates are 3 and 1 hr per unit for X and Y, respectively. Then, as before,

$$2X + 5Y = Z_{max} \qquad \text{(Eq. 1)}$$

$$2X + 1Y \leq 45 \qquad \text{(Eq. 2)}$$

In addition the firm is committed to a weekly purchase of at least 20 tons of a raw material produced by an affiliated plant, and product Y consumes twice as much of this as X. Then,

$$X + 2Y \geq 20 \qquad \text{(Eq. 3)}$$

A new element, involving an "equal to or greater than" sign has thus been introduced. The plant also produces 30 tons of a by-product from another process, all of which is to be used up, and both X and Y consume equal amounts of this ingredient, so that

$$X + Y = 30 \qquad \text{(Eq. 4)}$$

Our final restriction is thus in the form of an equality. In

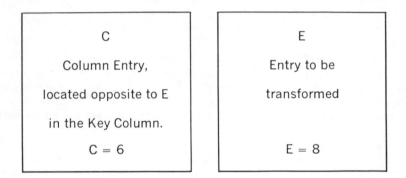

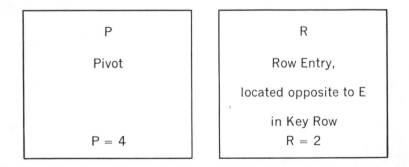

FIG. 14-2 Identification of matrix entries. In order to transform one matrix into another, we must first identify each entry in terms of C, E, P, and R as shown above. The illustrative example refers to the initial matrix of Fig. 14-1, using P = 4 as the Pivot and E ≠ 8 as the entry that is to be transformed. The actual transformation procedure then follows the simple formula given in Fig. 14-3.

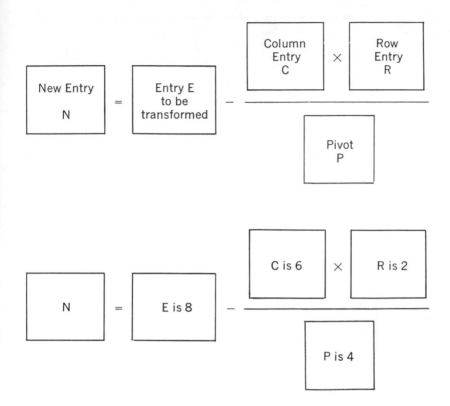

FIG. 14-3 Matrix transformation formula. Example shows application of formula in transforming Entry 8 of row M, column B of the first matrix. Note that C is 6, R is 2 and P is 4, so that we obtain the New Entry for the next matrix from $8 - (6 \times 2)/4 = 8 - 3 = 5$. This 5 is then utilized in place of E in the new matrix.

the following we will discuss the treatment of this extended model form.

Modified Equations

As before we must get rid of the inequality signs. For Eq. 2, we simply proceed as before, utilizing a slack variable S_1:

$$3X + Y + 1S_1 = 45 \qquad \text{(Eq. 2a)}$$

Recall that an equal sign implies that the left-hand side of a function equals the right-hand side. A proper value of S_1, therefore, somewhere between 0 or greater, should indeed

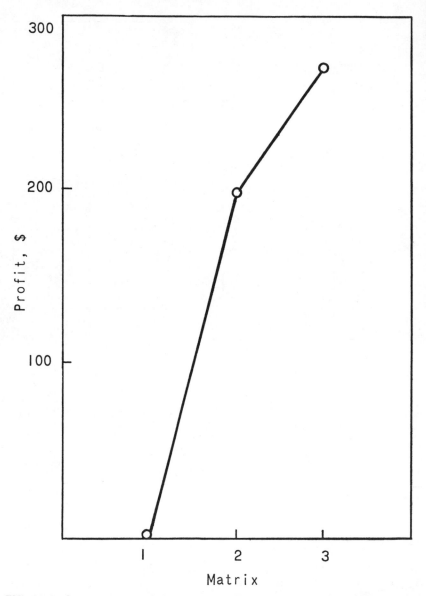

FIG. 14-4 Systematic search for optimum. The maximum profit of $272 per week was found in three successive matrix stages, beginning with an initial zero dollars, to $200 in the second stage and $272 in the final matrix.

raise the left-hand part of Eq. 2 so that instead of being "equal to or less than" the right-hand side, it becomes exactly equal to it.

Applying parallel reasoning to the "equal to or greater than" sign of Eq. 3 calls for a negative slack, thus:

$$X + 2Y - 1S_2 = 20 \qquad \text{(Eq. 3a)}$$

This form is not yet adequate, since the Simplex method of solving programming problems requires a matrix containing an "identity" portion. An Identity is a square block of data containing *positive* ones or "unity" in its diagonal and zeroes elsewhere. In order to meet this need, we shall now invent an artificial variable U, such that

$$X + 2Y - 1S_2 + 1U_2 = 20 \qquad \text{(Eq. 3b)}$$

The subscript 2 was arbitrarily selected for consistency with the slack subscript. We can justify our fictional U by simply considering the slack large enough in the negative direction to accommodate U's inclusion on the left–hand side of the equation. A further U_3 is invented for Eq. 4, giving

$$X + Y + 1U_3 = 30 \qquad \text{(Eq. 4a)}$$

where U_3 obviously must be considered zero-valued. The objective equation now becomes

$$2X + 5Y + 0S_1 - MU_2 + 0S_2 - MU_3 = Z_{\max} \qquad \text{(Eq. 1a)}$$

where the M represents a very large number, say $1,000, and the minus signs indicate that making the artificials will result in $1,000 loss per unit. This device assures us that the solution will never recommend production of any U's. The zero profits associated with the S's will similarly serve to indicate in the final solution that only the real products, X and Y, are profitable. Having completed these manipulations, we are ready to cast the four functions into symmetric form ready for matrix operations.

Symmetric Form

A little rearrangement of the equations leads to the symmetrical sets shown below. Each row contains all of the problem variables, but in many instances they are zero-valued; that is, we have combined them with a coefficient of zero, so that the effect of the variable vanishes.

$$2X + 5Y + 0S_2 + 0S_1 - MU_2 - MU_3 = Z_{max} \qquad \text{(Eq. 1b)}$$

$$3X + 1Y + 0S_2 + 1S_1 + 0U_2 + 0U_3 = 45 \qquad \text{(Eq. 2b)}$$

$$1X + 2Y - 1S_2 + 0S_1 + 1U_2 + 0U_3 = 20 \qquad \text{(Eq. 3c)}$$

$$1X + 1Y + 0S_2 + 0S_1 + 0U_2 + 1U_3 = 30 \qquad \text{(Eq. 4b)}$$

Let us rewrite the left-hand sides of the three restrictive equations without their variables:

$$
\begin{array}{cccccc}
3 & 1 & 0 & 1 & 0 & 0 \\
1 & 2 & -1 & 0 & 1 & 0 \\
1 & 1 & 0 & 0 & 0 & 1
\end{array}
$$

The right-hand segment, containing ones in the diagonal and zeroes elsewhere, forming a three-by-three square block of numbers, is known as the identity. The utilization of U_2 and U_3 in the formulation of the restrictive equations was dictated by our need to have this identity submatrix. The left-hand part of the matrix above is the so-called "basis." We are now ready to insert the data in standard Simplex matrix form and to solve by repeated matrix steps, as shown in the tableau of Fig. 14-5. The recommendation of the solution is to make 30 units of Y only, with a profit of \$150 per week.

Calculation Check

Finding an error in one of the initial rows, once a long series of calculations has been completed, can be quite annoying. These frustrations can be avoided by use of the check column in Fig. 14-5. In each row the check entry is the sum of all variables plus the entry under "Result." For example, in the first calculation row (b), we find: $1 + 2 - 1 + 1 + 20 = 23$. But there is more to the check. In particular, when we come to the first iteration, the old entry 23 divided by the pivot yields 11.5. This value occurs also in row h, and similarly the cross-addition of $0.5 + 1 = 0.5 + 0.5 + 10$ yields 11.5, thereby completing our check. For row c, the usual pivoting steps are

$$\text{New entry} = 50 - (23 \times 1)/2 = 50 - 11.5 = 38.5$$

	1	2	3	4	5	6	7	8	9	10	11	12
	Row	Coeff., C	Variables						Result	Check	Evaluation	Key row
			X	Y	S_2	U_2	S_1	U_3				
a		C →	2	5	0	$-M$	0	$-M$				↓
b	U_1	$-M$	1	2	-1	1	0	0	20	23	10	
c	S_1	0	3	1	0	0	1	0	45	50	45	
d	U_3	$-M$	1	1	0	0	0	1	30	33	33	
e	Z		$-2M$	$-3M$	M	0	0	0				
f	Z-C		$-2M-2$	$-3M-5$	M	$+M$	0	$+M$				
g		Key column	↑									
	First Iteration											
h	Y	5	1/2	1	$-1/2$	1/2	0	0	10	11.5	-50	↓
i	S_1	0	2.5	0	1/2	$-1/2$	1	0	35	38.5	70	
j	U_3	$-M$	1/2	0	1/2	$-1/2$	0	1	20	21.5	40	
k	Z		$-M/2$	0	$(-M-5)/2$	$(M+5)/2$	0	$-M$				
l	Z-C		$(-M-1)/2$	0	$(-M-5)/2$	$(3M+5)/2$	0	0				
m		Key column		←								
	Final Solution											
n	Y	5	1	1	0	0	0	1	30	33		
o	S_1	0	2	0	0	0	1	-1	15	17		
p	S_2	0	1	0	1	-1	0	2	40	43		
q	Z		5	5	0	0	0	5	150			
r	Z-C		3	0	0	M	0	$5+M$				

FIG. 14-5   Maximization problem solved through matrix steps, Simplex formulation.

	1	2	3	4	5	6	7	8	9	10	11	12
	Row	Coeff. C $\;\downarrow\;\rightarrow$	Variables						Result	Check	Evaluation	Key Row
			X	Y	S_2	U_2	S_1	U_3				
	(c_j)		2	5	0	M	0	M				
a												
b	U_1	M	1	2	−1	1	0	0	20	23	10	↓
c	S_1	0	3	1	0	0	1	0	45	50	45	
d	U_3	M	1	1	0	0	0	1	30	33	30	
e	Z		2M	3M	−M	M	0	M				
f	Z−C		2M − 2	3M − 5	−M	0	0	0				
g				←							Key column	

First Iteration

	1	2	3	4	5	6	7	8	9	10	11	12
	Row	Coeff. C	X	Y	S_2	U_2	S_1	U_3	Result	Check	Evaluation	Key Row
h	Y	5	$\tfrac{1}{2}$	1	$-\tfrac{1}{2}$	$\tfrac{1}{2}$	0	0	10	11.5	20	↓
i	S_1	0	$2\tfrac{1}{2}$	0	$\tfrac{1}{2}$	$-\tfrac{1}{2}$	1	0	35	38.5	87.5	
j	U_3	M	$\tfrac{1}{2}$	0	$\tfrac{1}{2}$	$-\tfrac{1}{2}$	0	1	20	21.5	40	
k	Z		$(M+5)/2$	5	$(M+5)/2$	$(5-M)/2$	0	M				
l	Z−C		$.5M+\tfrac{1}{2}$	0	$(M+5)/2$	$(5-3M)/2$	0	0				
m			←								Key column	

Second Iteration

n	Y	5	0	1	-0.6	0.6	-1.5	0	3	.38	-5 ↓
o	X	2	1	0	-1/5	-1/5	0.4	1	14	.154	-70
p	U_3	M	0	0	0.4	-0.4	-1/5	0	13	1.06	32.5
q	Z		2	5	$.4M + 2.3$	$2.6 - .4M$	$-2M - .2$	2			
r	Z-C		0	0	$.4M - 2.6$ ←	$2.6 - 1.4M$	$-2M - .2$	$2 - M$			
s		Key column									

Final Solution

t	Y	5	0	1	0	$-\frac{1}{2}$	1.5	22.5	24.5
u	X	2	1	0	0	$\frac{1}{2}$	$-\frac{1}{2}$	7.5	8.5
v	S_2	0	0	0	-1	$-\frac{1}{2}$	2.5	32.5	34.5
w	Z		2	5	0	$-3/2$	6.5	127.5	
x	Z-C		0	0	$-M$	$-3/2$	$6.5 - M$		

FIG. 14-6 Minimization problem solved in successive matrix steps.

Again, the addition of row i agrees, since $2.5 - 0.5 + 0.5 + 1 + 35 = 38.5$, so that both the row-total and the coinciding result of the pivotal condensation above assure us that all entries were obtained correctly (barring only the unfortunate possibility of two compensating errors).

Minimization Procedure

Suppose the values of $2 and $5 for products X and Y had represented costs instead of profits, and our goal were to minimize costs. The right-hand side of Eqs. 1 and 1b would now read Z_{min}. Next, to adjust for this change in direction, from maximum-seeking to minimum-searching, we multiply the left-hand side by -1. This mathematical manipulation reflects our reversal in objective, yielding

$$-2X - 5Y + 0S_2 + 0S_1 + MU_2 + MU_3 = Z_{min} \qquad \text{(Eq. 1c)}$$

The zero-valued slacks have remained positive despite the negative multiplication since, by convention, zero is always plus. The usual iterations, as given in Fig. 14-6, will yield a solution. This says: "To minimize costs under the factors and conditions of the problem, produce 22.5 units of product Y and 7.5 of product X. Resultant costs will be $127.50 per week."

Geometric Interpretation

The matrix steps may now be examined in the light of their geometric equivalents (Fig. 14-7).

The highest-profit point of $150 and the lowest cost point of $127.50 both appear on the graph at different corners. Other corners, corresponding to earlier matrix iteration steps, will also be apparent. For example, the preliminary recommendation (nonoptimal) to make 10 units of Y, given by the first iteration of the maximizing problem, has its geometric equivalent at the intersection of Eq. 3 with the Y-scale. Geometrically, too, this is a feasible but not optimal point.

As a further illustration of the parallel manner in which the matrix and geometric processes work, we may examine the minimization problem. The matrix steps, as a result of the second iteration, called for the production and sale of 3 units of Y and 14 units of X, which corresponds to the intersection of the lines for Eqs. 2 and 3 on the graphic solution. Again, of

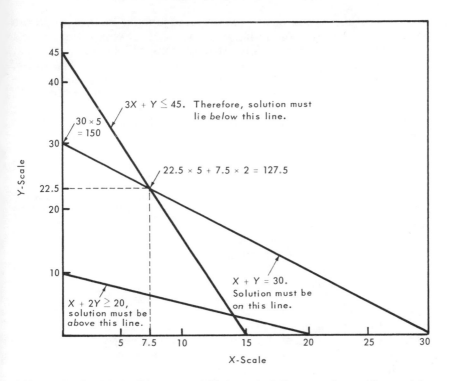

FIG. 14-7   Graphic interpretation of Mathematical Programming problem and its solution. Solution of maximization problem is at X = 0 and Y = 30 yielding 150. Solution of minimization problem is at X = 7.5 and Y = 22.5, yielding 127.5.

course, we are dealing with a point that contains a pre–optimum result.

Drawing a generalization from this illustration, it may be noted that in all programming problems involving linear but multidimensional relationships, each one of the matrix iteration steps has a corresponding geometric corner point. If it were possible to draw multidimensional graphs beyond the two-dimensional (or, isometrically, three-dimensional) limitations of paper, then the need for matrix calculations would cease.

Systematic Search Process

Stepwise pursuit of an optimum solution represents still another way in which the successive matrix procedures may be viewed. The three matrices of Fig. 14-1 will serve for illustration.

The initial matrix calls for output of just the imaginary items M and N, thereby exhausting capacity and leaving a total profit of zero. Obviously improvement will come by replacing at least some of the imaginaries with real products A and B. The trick, particularly in the solution of large-scale problems, is to move in that direction which will lead most quickly to an ultimate maximum profit.

Clues toward improvement are gleaned from an understanding of the meaning of rows M and N. Thus the first row, under column A, tells the manufacturer: "By giving up 6 hr of machining of M, you will gain enough capacity to make 1 unit of product A." Adjoining 6, the entry 8 indicates similarly that 8 hr of M must be relinquished for each unit of B. The one (1) under column M merely indicates that product M is a substitute for itself. For each hour of production of M given up, one new hour to produce M is gained. Finally, the zero under N means that N requires no machining, so no output of M need be given up. For row N, parallel interpretations apply, but this time with regard to assembly capacity.

Continuing our analysis, row Z-C shows for each product the loss per unit because the item is not made. For example, \$10 is lost for each unit of product A not made. Since the latter represents the highest loss in that row, it appears that profit will be improved most quickly if we replace some items M and N with A. How much shall this be, though?

The answer is held by the Evaluation column, which may also be viewed as representing the capacity in units for output of product A in machining and assembly. With 240 hr of machining time available, a rate of 6 hr per unit gives a capacity of 40 units. For assembly, 80 available hr divided by 4 hr per unit yields a capacity of 20 units. Since assembly is the bottleneck, we will study the changes needed to permit production of 20 units of A, utilizing a second matrix for this purpose.

In the new matrix we replace row designation N with A, since the assembly capacity for the latter will have to be won by relinquishing an appropriate amount of product N. Concomitantly the former zero profit is replaced by \$10 per unit. Also, recall that N had been presented as a whole, single unit. To do the same for A now, we divide all entries of the former row N by the amount in A, of 4. This gives 1 for the new row A and correspondingly quartered values for all remaining entries. We are well justified in this procedure, if it is remembered that the row entries are merely coefficients of an equation, and division by a constant is always permissible so long as all parts of the equation are so treated.

We have substituted A for N in assembly. But unlike the

latter, the former product also requires machining time. Some of the present capacity in that process must therefore be relinquished to accommodate A. Each entry in row M requires a reduction, the exact amount of which is reflected by the formula in Fig. 14-3. Note that this adjustment represents the proportion of capacity required to accommodate A. The Results column will serve as an example. For row M, the transformation is given by $240 - (6 \times 80/4)$, where 240 is the current time consumed in machining product M. The ratio $80/4$ gives the 20 units of A to be made, while the 6 constitutes the units of M that must be given up for each unit of A that is machined. Next, 6 times 20 gives 120, which subtracted from the original entry 240 gives the new value 120.

Substitution of A for N in the second matrix raises profit to \$200, but we know that an optimum has not yet been reached, since imaginary product M utilizes capacity to the extent of costing a \$2 loss per unit of potential output of B. At least one more matrix is needed to determine the proper amount of B to make. As it turns out, 24 units of B are required for optimum profit, and to produce this quantity our former tentative quantity of 20 units of A must be reduced to 8. The matrix steps thus represent a systematic sequence of successive adjustments and readjustments of quantity allocations until an optimum has been located. It is important to note, however, that these repetitive realignments do not interfere with actual output. Rather, they proceed on paper or on a computer, until such time as the solution found can be evaluated in practical terms and introduced in actuality.

Dynamic Decision Criteria, DDC

In Chap. 5, we extended MP procedures to include DDCs, and we shall now utilize the illustration given at that time to demonstrate how these criteria were derived (Fig. 14-8). The following minor changes were made:

1. Products X, Y, and Z were relabeled A, B, and C to avoid confusion with the Z-row.
2. The profits are shown in tens of dollars, so that \$90 becomes \$9, thereby simplifying the mathematics. The 10 can readily be restored by multiplying the profit figure of the solution (viz., Z_{max}).
3. The matrix calculations, being carried to many decimal points, show a slight (insignificant) difference in Z_{max}, (\$190.18 $\times$ 10 as against \$1905 originally).

Row	Coefficient	A	B	C	S_1	S_2	S_3	S_4	Result	Evaluation
		9	6	5	0	0	0			
Initial Matrix										
S_1	0	6	18	30	0	1	0		600	100
S_2	0	8	28	4	0	0	1		400	50
S_3	0	12	6	4	1	0	0		200	16.7
Z		0	0	0	0	0	0		0	
Z-C		-9	-6	-5	0	0	0			
Solution Matrix for Original Problem										
A	9	1	0	0	.0982	-.0112	-.0014		7.362	
C	5	0	0	1	-.0031	.0368	-.0230		12.2699	
B	6	0	1	0	-.2076	-.0020	.0429		10.4294	
Z		9	6	5	.7027	.0712	.1298		190.18	
Z-C		0	0	0	.7027	.0712	.1298			
Modified Initial Matrix for Dynamic Decision Criterion Applying to Product A										
New Coeff.		1	0	0	0	0	0	0		
A	1	1	0	0	.0982	-.0112	-.0014	0	7.362	-657
C	0	0	0	1	-.0031	.0368	-.0230	0	12.2699	333
B	0	0	1	0	-.0276	-.0020	.0429	0	10.4294	-5215
S_4	0	0	0	0	.7027	.0712	.1298	1	9.5090	134
Z		1	0	0	.0982	-.0012	-.0014	0	7.3620	
Z-C		0	0	0	.0982	-.0112	-.0014	0		
Solution Matrix with Dynamic Criterion for A										
A	1	1	0	0	.2087	0	.0190	.1573	8.8578	
C	0	0	0	1	-.3663	0	-.0901	-.5169	7.3551	
B	0	0	1	0	-.0079	0	.0465	.0281	10.6965	
S_2	0	0	0	0	9.8694	1	1.8230	14.0449	133.553	
Z		1	0	0	.2087	0	.0190	.1573	8.8578	
Z-C		0	0	0	.2087	0	.0190	.1573		

FIG. 14-8 Original problem and solution, followed by determination of Dynamic Decision Criterion.

The initial matrix and its solution appear in the first two tableaus, yielding the conventional MP solution. For the purpose of a 95% DDC, we are willing to take a total loss of 5%, or 0.05 × $190.18 = $9.509 in relation to the optimum. Since the entries in the Z-C row indicate the minimum cost of deviating from the optimum by one unit for each of the products A through S_3, it is apparent that our 95% DDC in equation form is

$$0A + 0B + 0Z + 0.7027\,S_1 + 0.0712\,S_2 + 0.1298\,S_3 \leq 9.509$$

An equal sign for this expression is obtained through the addition of another slack, $1S_4$. The resultant equation is entered in row S_4 of the modified initial matrix. In addition, the coefficients for all of the variables become zero; excepting for the product for which a DDC is to be found (A in our example), where the coefficient becomes 1.

Matrix solution of the modified tableau, which in this instance requires only one iteration, gives a modified value of 8.8578 for product A. Subtracting the original 7.362 yields the DDC of 1.5. We interpret this result as previously discussed in Chap. 5. Upon a review of the details in Fig. 14-8, the reader will be enabled to check for himself the validity of the DDCs for products B and C.

Summary

Although an understanding of this chapter is not essential for practice-oriented executives, nevertheless it serves to emphasize the flexibility and resourcefulness of mathematical methods for dealing with a variety of situations.

As all mathematical tools, MP is far from being a panacea. Moreover, MP experts would criticze our discussion for having been confined to linear terms, while real-world problems often involve nonlinear objectives or restrictions. We must not overlook the fact, however, that even for real-world problems it becomes exceedingly difficult to establish the nature of nonlinear functions. For example, there is little doubt that for many items market demand changes in nonlinear relation to increases or decreases in price. We could program a curvilinear demand function, but the trouble is that management usually has little information on just how the curve runs. A practical alternative is to assume linearity for small segments of the curve and program accordingly. The real world, not the mathematics, forces us to proceed in this approximate manner.

Finally, MP is not substituting its findings for managerial judgment. A serious error, resulting from failure to consider nonlinearities, should be caught by the managers reviewing the program.

PART II

Supplemental Case—Histories
of Solved Problems

15

CASES PERMITTING GRAPHIC
SOLUTION

The case histories provided here are designed to serve those readers who wish to enhance their proficiency in the recognition, handling and solution of Mathematical Programming problems amenable to graphic methods.

The material is so arranged that the case-problem is presented first. The approach used to yield a solution is next given, together with the results obtained. Self-study is thus encouraged. The reader-student can first attempt his own solution. Most of the time he is likely to be successful. Where he fails, a turning of the page brings the procedure leading to a solution.

CASE A-1

Claxton and Horn Associates:
Investment Planning

Because earnings of this industrial consulting firm tended to fluctuate considerably from year to year, management had decided to invest some of the firm's funds in stocks rather than in expansion of business facilities. Two stocks were viewed with particular favor:

	Allstone Corp.	Balldine Inc.
Acquisition cost, $/share	10	8
Expected annual gains for the next five years, $/share:		
Cash dividends	10	10
Stock dividends	8	12
Capital appreciation	12	6

Since cash dividends tended to be relatively stable for both stocks, the partners desired to invest so as to receive at least $1000 per year from this source. Each year both corporations had been issuing stock dividends, and the partners wished to aim at $960 per year from this source. Capital gains were less certain, but nevertheless they wanted to try for at least $720 per year in capital appreciation. Naturally they desired to accomplish their aim with a minimal investment.

Required

How many shares of each stock should they buy and what will be the total cost of the purchase?

SOLUTION TO CLAXTON AND HORN'S INVESTMENT PROBLEM

In order to attain the desired investment income at minimum investment expenditure, 80 shares of Balldine (at $8 per share) and 20 shares of Allstone (at $10 per share) should be purchased. The results is a total cost of $640 and 200 or $840.

Graphic analysis of the problem is adequate, as shown by Fig. A-1.

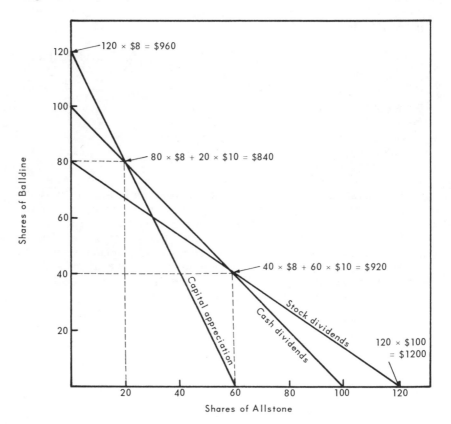

FIG. A-1   Claxton and Horn Associates, graphic solution. Feasible solutions are shown by the lines bordering to the right-hand side, such as 120 shares of Balldine, 80 Balldine and 20 Allstone, 40 Balldine and 60 Allstone, and 120 Allstone. Of these, 80 shares of Balldine at $8 a share and 20 shares of Allstone at $10 per share, total-ing $640 + $200 = $840, represent the lowest-cost solution.

Denoting the number of Allstone shares by A and Balldine by B, the objective equation becomes

$$10A + 8B = Z_{min}$$

For the restrictive equations, we have

$$10A + 10B \geq 1000$$

$$8A + 12B \geq 960$$

$$12A + 6B \geq 720$$

The matrix steps of the Simplex method will give an identical result.

Based on the solution obtained, new questions may be raised. For example, 120 shares of Allstone stock are feasible at an investment of $1200, which is above the minimum of $840, but the income will be considerably greater. Therefore the original problem might be reviewed by the partners with a view to formulating a new one: What investment will maximize income regardless of source? In practice it is often found that the solution of one MP problem brings up new questions that lead to revised question-posing, new and more sophisticated problems and thus deeper insights and better data for decision-making.

CASE A-2

Precisioneer Machine Shops: Quality Control Inspection Schedules

Management felt that extensive inspection was needed, because the high precision close-tolerance work produced on their equipment had to meet such demanding specifications that 10% of the product was defective prior to inspection. The following are the pertinent data:

	(1) Supervisory inspector	(2) Staff inspector	(3) Total
a. Weekly pay, $	300	200	
b. Lot quality before inspection, % defective			10
c. Inspector proficiency, %	98	96	
d. Inspection rate, pieces per week	2000	1500	
e. Cost of missed defective piece, $			$10
f. Minimum number of pieces to be inspected per week			12,000

The term "supervisory inspector" is somewhat of a euphemism. The supervisory inspector is also responsible for inspection work (2000 pieces per week), but he is paid an hour overtime at the end of each day to fill in inspection reports and be available for quality control meetings. Because of these additional responsibilities and because of the higher caliber of work of the supervisory inspector, his weekly pay rate is also better. Management felt that there should be two staff inspectors per supervisory inspector and that the latter should be responsible for the inspection reports for all three of them.

Required

1. What is the minimum number of inspectors of each type that we must employ?

2. What will be the weekly inspection cost?
3. Express the problem in equation form.

This problem can be solved geometrically or, if preferred, by the matrix steps of the Simplex procedure.

SOLUTION TO PRECISIONEER'S INSPECTION PROBLEM

The problem is analyzed in Table A-2. The optimum number of inspectors is three supervisors and six staff, with a weekly cost of $2580. Geometric determinations appear in Fig. A-2.

Table A-2 Inspection Problem and Its Solution

Problem data and solution*	(1) Supervisory inspector	(2) Staff inspector	(3) Total
a. Weekly pay, $	300	200	
b. Lot quality *before* inspection, % defective pieces per lot			10
c. Inspector proficiency, %	98	96	
d. Inspection rate, pieces per week	2000	1500	
e. Cost of a missed defective piece, $			$10
f. Minimum number of pieces to be inspected per week			12,000
g. Inspection error, % $= 100\% - c\%$	2	4	
h. Lot quality *after* inspection, % defective pieces per lot $= b\% \times g\% = 0.10 \times g\%$	0.2	0.4	
i. Weekly cost of unremoved defectives, $ $= h\% \times e \times d = h\% \times \$10 \times d$	40	60	
j. Combined weekly cost of pay and unremoved defectives, $ $= a + i$	340	260	
k. No. of inspectors needed if only supervisors *or* staff inspectors but not both can be employed $(= 12,000/d)$ (These points are needed for plotting the graph, Fig. A-2).	6	8	
l. Optimum inspectors (from graph)	3	6	
m. Weekly total inspection cost, $ $= l \times j$	1020	1560	2,580

*Problem data in lines *a* to *f*, derived data leading to solution in lines *g* to *m*.

Using symbols A and B for supervisory and staff inspectors, respectively, and the cost figures of $340 and 260 per week for each, we obtain

$$340A + 260B = Z_{min} \quad \text{(objective equation)}$$
$$2000A + 1500B \geq 12,000 \quad \text{(first requirement)}$$
$$2A = 1B \quad \text{(additional specification)}$$

Since fractional inspectors do not exist, the solution must be in terms of integers.

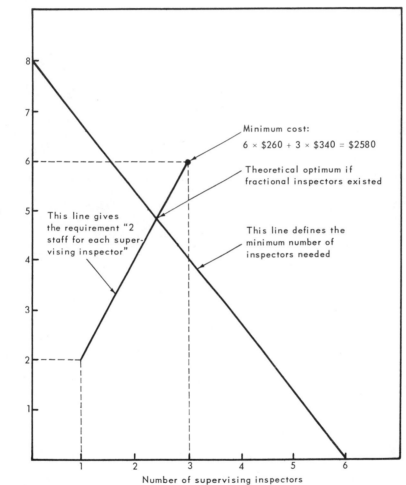

FIG. A-2 Minimum-cost inspection set-up. The solution is to use three supervising inspectors and 6 staff inspectors.

CASE A-3

Delta Tank Operation: Strategy Development

The commander of a small tank group has been ordered to win and occupy a valley located in a river delta area. He has the following crews, equipment, and strategic data:

	Heavy tank	Light tank	Total
a. Number of tanks available	4	10	
b. Men needed per tank	4	2	
c. Total number of men available			24

Although the fire power of the heavy tank is three times that of the light tank, yet the commander felt that he should use more light tanks than heavy ones, since the latter were more effective against guerillas.

Required

1. How many tanks of each type should the commander send into combat, keeping in mind that he wishes to maximize total fire power within the limitations stated above?
2. If you could persuade the commander that the rule "more light tanks than heavy" may not be applicable to the present situation—reconnaisance has indicated the presence of merely regular enemy forces and no evidence of guerillas—then how many tanks of each type could be used?
3. How many men are used in (1) and (2) above?
4. Support your findings with geometric or matrix analysis, as may be appropriate.

SOLUTION TO DELTA TANK
OPERATION PROBLEM

A geometric analysis of the strategy, in Fig. A-3, yields these results:

1. Maximal fire power is accomplished, within the tank-ratio restrictions, by using three heavy and six light tanks with a fire power rating of 15.
2. If the tank-ratio restriction is removed, we can send four tanks of each kind, with a fire power rating of 16.
3. Manpower requirements are 24 men under both of the above two programs.

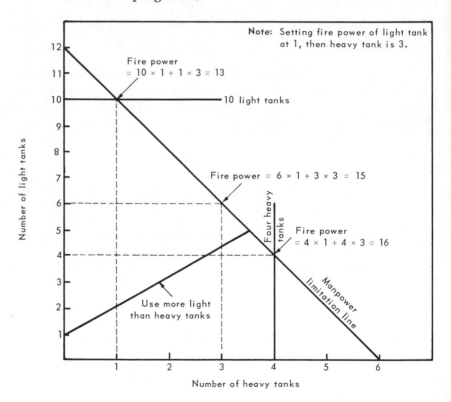

FIG. A-3   Tank strategy optimization. The theoretical optimum of 4 heavy and 4 light tanks, with a total fire power of 16 (assuming a light tank has a power of 1, so that a heavy tank is worth 3), does not meet the requirement that light tanks must exceed heavy ones. The next integer, 3 heavy tanks, permits 6 light ones with a fire power of 15. No further points need to be checked, since higher portions on the manpower limitation line lead to lower fire power.

4. Matrix analysis would lead to identical results. Using the symbols H and L for heavy and light tanks, we obtain

$$3H + 1L = Z_{max} \quad \text{(objective equation)}$$

$$4H + 2L \leq 24 \quad \text{(manpower restriction)}$$

$$L \leq 10 \quad \text{(light tank limit)}$$

$$H \leq 4 \quad \text{(heavy tank limit)}$$

$$L > H \quad \text{(strategic opinion factor)}$$

The last equation is equivalent to writing

$$L + 1 \leq H$$

since fractional tanks are impossible.

With the tank-ratio restriction removed, the last equation above disappears.

CASE A-4

Lawrencetown Distillers: Optimal Blending of Ingredients

Production of alcohol involves the distillation of a fermented mash, resulting in a liquid residue or "stillage," which upon drying and other treatments becomes a valuable feed. Lawrencetown was marketing a high-quality feed mix of guaranteed vitamin constituents. Requirements, production factors, and market conditions appear below:

	Own dried stillage	Purchased additive	Total
a. Cost per pound, $	0.20	2.00	
b. Usual amount of vitamin units per pound of feed mix			
A	1	15	
B	2	10	
C	6	50	
c. Minimum units of vitamin to be contained in a mixer load			
A			600
B			800
C			3000
d. Productive capacity in pounds per month	18,000		
e. Market supplies available per month, lb		2,400	
f. Number of mixer loads to be produced per month			30

Required

Determine the minimum-cost blending proportions per mixer load in terms of dried stillage and purchased additive. What is the total dollar cost involved?

SOLUTION TO LAWRENCETOWN'S BLENDING PROBLEM

As a first step, a tabular analysis is made, which provides the data points used for the geometric interpretation in Fig. A-4.

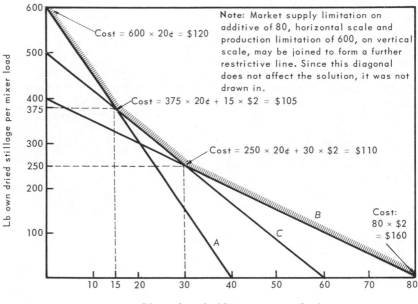

FIG. A-4   Feed mix blending problem. Shaded line segments identify feasible solutions. Minimum cost corresponds to 15 lb of purchased and 375 lb of own materials, at $105 per mixer load.

Minimum cost of $210 per mixer load is accomplished by utilizing 250 lb of our own dried stillage and 30 lb of additive.

In equation form, the following is found, with symbols D and P referring to dried stillage and purchased additive, respectively:

$$0.20D + 2.00P = Z_{min} \quad \text{(objective equation)}$$

$$A \geq 600 \qquad \text{(vitamin A requirement)}$$

$$B \geq 800 \qquad \text{(vitamin B requirement)}$$

$$C \geq 3000 \qquad \text{(vitamin C requirement)}$$

$D \leq 18,000$ (capacity limitation)

$P \leq 2400$ (market availability limitation)

As noted from the graphic analysis, the capacity limitations and market availability limitations are less restrictive than the vitamin requirements. Accordingly the latter govern the area within which a solution occurs.

Table A-4 Lawrencetown's Blending Problem and Its Solution

	(1) Own dried stillage	(2) Purchased additive	(3) Total
a. Cost per pound, $	0.20	2.00	
b. Vitamin content in units per pound of feed mix			
A	1	15	
B	2	10	
C	6	50	
c. Minimum vitamin units needed per mixer load			
A			600
B			800
C			3000
d. Capacity and availability per month	18,000	2400	
e. Mixer loads per month			30
f. Capacity and availability per mixer load = d/e	600	80	
g. Amount of feed mix needed per mixer load if only dried stillage _or_ purchased additive, but not both, are used (= c/b)			
A	600	40	
B	400	80	
C	500	60	
h. Minimum cost feed mix (from geometric analysis, using data points found in "g").	375	15	
i. Cost (= a × h)	75	30	105

CASE A-5

Finespun Worsted Mill: Determining
Optimal Work Assignments and Layouts

Staple yarns are spun from a relatively coarse, heavy strand of fibers about an inch in diameter. In successive processing stages, attenuation and drafting occurs, whereby this coarse strand is reduced to a fine yarn. One of the earlier drafting stages is known as "pin-drafting." An initial "breaker" operation is followed by a "finisher" drawing machine. By changing various gears, pulleys, and guides on a machine, the combination of breakers and finishers can be varied on a layout without adversely affecting quality.

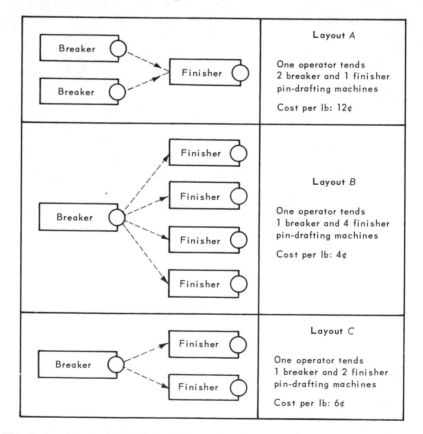

FIG. A–5a   Layouts of equal feasibility and producing similar qualities on a battery of pin–drafting machines.

Time studies, quality evaluations, and cost analyses revealed three possible layouts and workloads, as shown in Fig. A–5a, at costs per pound of output of 12, 4, and 6 ¢ for workloads *A*, *B*, and *C*, respectively. The mill had 12 breaker and 20 finisher machines and it desired to utilize all of them.

Required

1. Find the lowest-cost set-up and the number of pin-drafters of each type utilized by it.
2. How many operators will be required?

SOLUTION TO FINESPUN'S ASSIGNMENT AND LAYOUT DETERMINATION

Based on the analysis in Table A–5, the graphic solution in Fig. A–5b is obtained.

Table A-5 Optimum Layout Problem Analysis

Problem data and analysis	(1)	(2)	(3)	(4)
	Type of layout			Total
	A	B	C	
a. Cost per pound, ¢	12	4	6	
b. Assignment, machines per operator:				
Breakers	2	1	1	
Finishers	1	4	2	
c. Machines to be utilized:				
Breakers				12
Finishers				20
d. Assignments of machines possible if *only* A, *or* B *or* C workloads are used:				
Breakers (= 12 div. by b)	6*	12	12	
Finishers (= 20 div. by b)	20	5*	10*	
e. Optimum no. of layouts (from graphic solution)	4	4	–	
f. Machines utilized (= b × e):				
Breakers	8	4		12
Finishers	4	16		20
g. Operators utilized (= f/b)	4	4		
h. Total cost, ¢, (= a × e)	48	16		64†

*Not permissible, because breakers or finishers are under-utilized.

†Lowest total cost of running all machines.

Finespun should use four layouts of type A and 4 of type B, involving a total of 12 breakers and 20 finishers at an overall cost of 64 ¢ per pound.

In the graphic analysis, observe that the horizontal axis (abscissa) serves simultaneously as the scale for layouts B and C. In this way we saved ourselves the trouble of another graph. The reader might wonder why we did not use a graph with B on the vertical axis (ordinate) and C on the abscissa. The reason is that the lines for breakers and finishers would not meet on such a graph (try it!), and therefore we need not consider this combination.

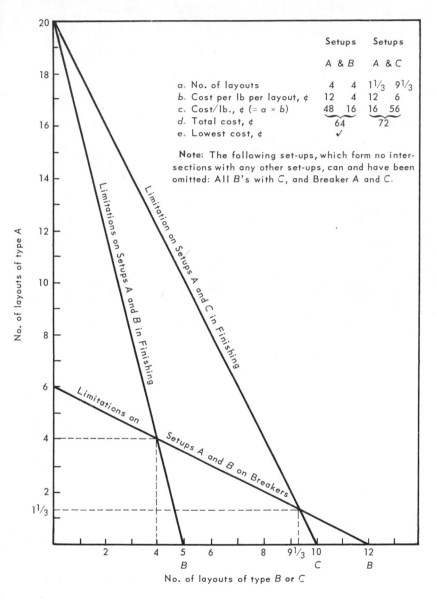

FIG. A–5b Optimum set-up. Graphic analysis shows that 4 Type A layouts and 4 Type B layouts will satisfy management's needs at the lowest cost.

Using symbols K and F for breaker and finisher pin drafters and maintaining the layout designations A, B, and C, we obtain:

$$12A + 4B + 6C = Z_{min} \quad \text{(objective equation)}$$
$$K = 12 \quad \text{(requirement for breakers)}$$
$$F = 12 \quad \text{(requirement for finishers)}$$

It is thus apparent that we are working with three variables in the objective equation, hence the need for the double-scale in the geometric analysis. An alternative might have been to construct a three-dimensional geometric model on isometric graph paper.

CASE A-6

Mirrorglaze Finishing Corporation: Planning of Commision Polishing and Finishing Operations

Mirrorglaze was in the business of doing polishing, finishing, and plating on a commission basis for a number of manufacturers, involving such diverse items as electronic components and holloware. Favorable market conditions assured an abundance of available business, particularly in the polishing and finishing lines. At present the firm had an offer to polish and finish three products, A, B, and C at prices that would yield $6, 5, and 4 profit per gross, respectively. Available uncommitted capacity, in hours per week, is 138.6 in polishing and 510 in finishing. Production rates are as follows:

	Product A	Product B	Product C
Polishing, hours per gross	9.9	7.7	11.55
Finishing	51	17	30

Naturally, the firm desires to accept only such business as will maximize profits.

Required

1. Determine the optimum volume of business to accept.
2. What is the weekly profit so obtained?

SOLUTION TO MIRRORGLAZE'S
POLISHING-FINISHING TASK

A tabular analysis supplemented by the graphs in Fig. A-6 indicates the following:

1. Accept a volume of 18 gross of product B per week for maximum profit.
2. Profit per will will be $90.00.

Observe that the program fully utilizes capacity available.

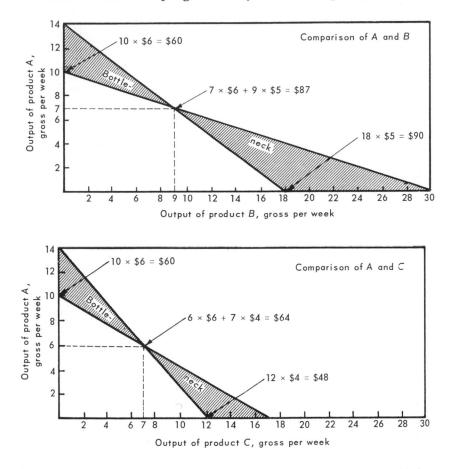

FIG. A-6   Analysis of three products and two processes for maximum yield. It is seen that the upper diagram yields maximum profit, if 18 gross of Product B only are made. Note that we did not compare Outputs of B and C, with B on one axis and C on the other. The reader may do this for himself, and he will note that the lines do not intersect; hence this comparison is not needed.

Table A-6   Polishing-Finishing Program Analysis

Problem data and solution	(1) Product A	(2) Product B	(3) Product C	(4) Total
a. Profit, \$/gross	6	5	4	
b. Production rate, hr per gross				
Polishing	9.9	7.7	11.55	
Finishing	51	17	30	
c. Available capacity, hr per week				
Polishing				138.6
Finishing				510.0
d. Production, gross per week, if only A or B or C is produced ($= c/b$)				
Polishing	(14)*	18	12	
Finishing	10	30	(17)	
e. Profit associated with production of only A or B or C ($= a \times d$), \$/week	60	90	48	
f. Combination program (from Fig. A-6)	7	9	0	
g. Production, hours per week, used by combination program				
Polishing ($= b \times f$)	69.3	69.3	0	138.6
Finishing ($= c \times f$)	357	153	0	510.0
h. Profit involved in combination program, \$ ($= a \times f$)	42	45		87
i. Maximum-profit program, \$/week ($=$ highest value from a check of rows e and h)		90†		90†

*Parentheses indicate bottleneck. For example, only 10 of A can be polished in the example in row d, because finishing is the limiting process.
†Total profit per week is at a maximum if *only* product B, 18 gross, are produced, yielding \$90.

Using symbols P and F for polishing and finishing and maintaining A, B, and C for the three products, we find:

$$6A + 5B + 4C = Z_{max} \qquad \text{(objective equation)}$$

$$9.9A + 7.7B + 11.55C \le 138.6 \quad \text{(polishing limitation)}$$

$$51A + 17B + 30C \le 510.0 \quad \text{(finishing restriction)}$$

This, again, is a three-variables problem. A three-dimensional graph is required, or else a set of three ordinary (two-dimensional) graphs may be considered. In the present instance, only two of the two-dimensionals are needed, because the graph with outputs B and C on its axes does not yield an intersecting point for the polishing and finishing lines. (Try it! This combination need not be considered, therefore.) Observe also that we could have combined the two graphs into one, simply by using a double-scale (as was done in the previous case illustration).

The problem also serves to demonstrate a general (mathematically provable) rule: The number of products or other items in solution can never exceed the number of restrictions. Therefore, although we had three variables in the objective equation, with only two restrictive equations only two of the three variables can come into the solution.

CASE A-7

Style-Rite Knitwear: Production-Sales Coordination

Style-Rite was primarily engaged in knitting worsted and synthetic yarn outerwear, produced in confined patterns to definite orders from chain stores and mail distribution houses. Because of a poor season, unused capacity remained, and management was considering the possibility of knitting for sale to smaller stores, two newly developed sweaters, called "lounger" and "swinger," respectively.

Based on planned prices, these two styles would yield profit contributions of $80 and $60 per gross, respectively. But it was not certain whether this price could actually be maintained, and sales might have to be at prices so low that no profit would be realized. After considerable discussion among the management people involved, particularly the mill manager, sales manager, controller, and president, it was the general consensus that there was only a 50% chance of attaining the aimed at profits quoted above. Therefore, applying this 50% to $80 and $60, respectively, gives expected profits of $40 and $30.

Since the two models utilized productive capacity in different degrees, the question then was in what proportions to seek to promote each style so as to maximize expected profits.

The company's controller undertook to gather together the pertinent data, which would be useful in the decision-making process. These included production time requirements, productive capacity, and an analysis of the output obtainable if only the Lounger or the Swinger but not both styles were made and sold. The data appear in Table A-7.

The analysis showed that only the Lounger should be produced at a profit of $400 per day. The mill manager was not so sure this represented the right solution, because it left a great deal of idle winding machinery. Sales was not impressed either, because making just one style limited the number of stores that would be interested in the merchandise.

Required

Make an analysis of this problem and report your findings.

Table A-7 Style-Rite's Coordination Problem

Problem data gathered	Styles		Total
	Lounger	Swinger	
a. Profit contribution per gross, $	40	30	
b. Production rates, hr per gross			
Cone winding	2	6	
Knitting	8	4	
c. Available uncommitted capacity, machine-hours per day			
Winding frames			60
Knitting machines			80
d. Output attainable if only the Lounger *or* the Swinger, but not both, are produced (= c/b)			
Winding	(30)	10	
Knitting	10	(20)	
(Parentheses indicate nonfeasible quantities because of bottleneck in the other process)			
e. Profit from either program in d, $ (= $a \times d$, using only the open d's)	400	300	
f. Optimal one-style program profit, $	400		400

SOLUTION TO STYLE-RITE'S
COORDINATION PROBLEM

A geometric analysis (Fig. A-7a) reveals that Syle-Rite's optimum profit of $480 per day is attainable, under present problem restraints, by producing 6 gross of the Lounger and 8 gross of the Swinger, daily. By utilizing capacity fully, this program does not leave any idle machinery and results in $80 per day more than the original analysis.

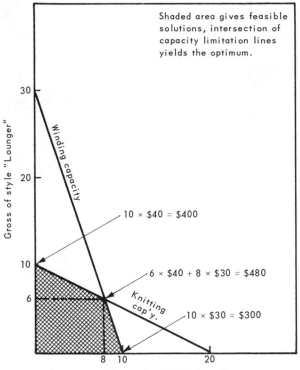

FIG. A-7a Analysis of Style-Rite's production and sales problem. Optimum is to produce 8 gross of Swinger and 6 gross of Lounger.

It is, of course, possible that the geometric analysis would have again favored just one and not both styles. Then, in order to meet the Sales Department's "diversified product line" desires, we would have to establish the minimum quantities of each style that the department needs. These data then form additional restrictions. An examination of the equations, using "L" for Lounger quantities and "G" for Swinger quantities, will demonstrate:

$$40L + 30G = Z_{max} \quad \text{(Style-Rite's objective)}$$

$$2L + 6G \leq 60 \quad \text{(Winding Limitation)}$$

$$8L + 4G \leq 80 \quad \text{(Knitting Limitation)}$$

The following further restrictions now apply, assuming that the Sales Department requires at least 3 gross per day of each style:

Row	Coefficient, C	L 40	G 30	S_1 0	S_2 0	Result	Evaluation	Check	Key
S_1	0	2	6	1	0	60	30	69	
S_2	0	⑧	4	0	1	80	10	93	←
Z		0	0	0	0	0			
Z-C		-40	-30	0	0				
Key		↑							
Second Matrix Tableau, First Iteration									
S_1	0	0	⑤	1	$-\frac{1}{4}$	40	8	183/4	←
L	40	1	$\frac{1}{2}$	0	1/8	10	20	93/8	
Z		40	20	0	5	400			
Z-C		0	-10	0	5				
Key			↑						
Third Matrix Tableau, Second Iteration									
G	30	0	1	1/5	-1/20	8		183/20	
L	40	1	0	-0.1	3/20	6		141/20	
Z		40	30	2	9/2	480			
Z-C		0	0	2	9/2				

FIG. A-7b   Matrix steps, Simplex formulation, leading to the identical results as the prior geometric analysis. Maximum profit of $480 per day (Third Matrix, Row Z) results from the production and sale of 8 gross of G and 6 gross of L. The intersection of key row and key column arrows points to the pivot.

$$L \geq 3 \quad \text{(Lounger minimum)}$$

$$G \geq 3 \quad \text{(Swinger minimum)}$$

Further product diversification, in terms of number and variety of styles may be desired; but it should be remembered that we are merely trying to sell excess capacity for short-term seasonal demands. Increased style proliferation will bring higher unit costs, and the gain in merchandising flexibility may be offset by rising production and administrative costs. The matrix steps in Fig. A-7b for the current problem can, of course, be readily expanded to accommodate many more potential styles; but the analysis can no longer be demonstrated to management in graphic form.

The validity of the MP analysis will depend, to a large extent, upon the managerial estimates of the likelihood of attaining various prices and thus profits. This use of "subjective probability" is justified, however, because obviously it is neither feasible nor practical to conduct objective studies of market potentials based on probability sampling surveys. It may be expected, however, that the managerial estimates are based at least in part on some consultation with prospective customers. Experience and judgment provide further basis for the subjective probabilities used in the problem formulation.*

---

*The reader interested in further applications of subjective probability in marketing management is referred to Frank J. Charvat and Tate W. Whitman, Marketing Management, A Quantitative Approach, Simmons-Boardman, New York, 1964.

16

CASES CALLING FOR SIMPLEX ANALYSIS

When large-scale problems require solution, graphic approaches become inadequate, and resort must be had to matrix methods. The most practical of these is the so-called "Simplex Algorithm" or "Simplex Method," as described in Part I of the book.

As was done for the graphic materials, we provide the problem first, thus giving the reader an opportunity to make his own solutions before turning the page.

CASE B-1

Flowmaster Incorporated: Optimum
Product Mix

Rapidly accelerating technologies had created a good demand for the high-precision pumps, produced under several patents, by Flowmaster. Recently, three new models, the Accu-Flow, the Brine-Proof, and the Calimete had been developed and test marketed. Management was assured that the market was considerable, but available productive capacities were limited. Even with overtime, three-shift production, only the following hours could be squeezed in per week: 200 in Machining, 600 in assembly, and 400 in reliability and serviceability testing.

Production rates and profit contributions per pump appear below:

	Accu-Flow	Brine-Proof	Calimete
Profit contribution, $/pump	9	6	5
Production rate, hr per pump			
Machining	12	6	4
Assembly	6	18	30
Testing	8	28	4

Management desired to know what product mix, that is, quantity per week of each model, would maximize profit.

Required

Perform the required analysis.

SOLUTION TO FLOWMASTER PROBLEM

Although geometric procedures can be used with some ingenuity, it is probably simpler and more accurate to proceed
via the Simplex method (Fig. B-1). The results are in decimals,
which may require rounding for practical purposes. Designating
the three pumps by A, B, and C (according to their starting

Row	Coeffi-cient	A	B	C	S_1	S_2	S_3	Result	Evalu-ation
		9	6	5	0	0	0		
S_1	0	12	6	4	1	0	0	200	16.7
S_2	0	6	18	30	0	1	0	600	100
S_3	0	8	28	4	0	0	1	400	50
Z		0	0	0	0	0	0		
Z-C		-9	-6	-5	0	0	0		
First Iteration									
A	9	1	$\frac{1}{2}$	1/3	1/12	0	0	16.7	50
S_2	0	0	15	28	$-\frac{1}{2}$	1	0	500	17.86
S_3	0	0	24	4/3	-2/3	0	1	266.7	200
Z		9	9/2	3	3/4	0	0	150	
Z-C		0	-3/2	-2	3/4	0	0		
Second Iteration									
A	9	1	9/28	0	5/56	-1/84	0	10.71	33.33
C	5	0	15/28	1	-1/56	1/28	0	17.86	33.33
S_3	0	0	163/7	0	-9/14	-1/21	1	242.9	10.43
Z		9/7	39/7	5	5/7	1/14	0	185.71	
Z-C		0	-3/7	0	5/7	1/14	0		
Third Iteration									
A	9	1	0	0	.1	-.01	.01	7.36	
C	5	0	0	1	-.01	.03	-.02	12.27	
B	6	0	1	0	-.03	-.002	.04	10.43	
Z		9	6	5	.67	.05	.23	190.17	

FIG. B-1   Simplex steps leading to solution of pump model problem. Shadings represent pivots. (Calculations due to Miss Gail P. Anderson, graduate student assistant.)

letters), the optimum program recommends 7.36 of A, 10.43 of
B, and 12.27 of C, with a weekly profit contribution of $190.17.
Equations are given below:

$$9A + 6B + 5C = Z_{max} \quad \text{(objective)}$$

$$12A + 6B + 4C \leq 200 \qquad \text{(machining capacity)}$$

$$6A + 18B + 30C \leq 600 \qquad \text{(assembly capacity)}$$

$$8A + 28B + 4C \leq 400 \qquad \text{(testing capacity)}$$

The inequalities convert readily to equal signs by the addition of three slacks, one for each of the capacity limitations.

Since market conditions are favorable, there is no reason why the firm cannot adopt this program and achieve full sale of all that is produced.

CASE B-2

Uniduct Electric Company: Profitability of Electric Wire Program

Recently the owner of this small firm had purchased three copper drawing machines which, together with his plastic extruder, represented his principal equipment. Thus the three machines drew the wire to the required gage. Next the required number of strands were processed through the extruder, which provided the plastic insulation. Both single and parallel conductor wire was produced. Because of the limited amount of equipment, only three operators could be employed.

Previously the owner had not had any problems in determining the optimum product mix, that is, which products in what quantities would yield maximum overall profit, based on expected prices and costs and the limitations set by the capacity of his equipment. He had had only one machine, the extruder, which was operational 900 hr per year. The rest of the time was utilized for setups, repairs, mandatory cooling-off periods, and preventive maintenance. The copper drawing equipment required much less downtime, so that the three machines pro-

Table B-2a   Electric Wire Manufacturing Program

	Products Saleable			
	A	B	C	D
a. Type conductor	Single	Single	Parallel	Parallel
b. Gage no.	18	22	18	18
c. No. of strands	16	7	41	41
d. Copper no.	30	30	34	34
e. Plastic insulation, thickness, in.	1/32	1/64	0.031	0.047
f. Profit $ per million ft	100	140	500	750
Production rate, hr per million ft				
g. Plastic extruder (allowing for downtime)	8	55	15	25
h. Copper drawing (3 machines)	60	40	400	400
i. Human labor (3 operators)	70	35	400	450

Source: Problem data adapted from an industrial case study reported by Michael A. Tirabassi in a term paper in the author's course "Systems Analysis and Design."

vided an annual capacity of 9500 hr. He estimated that the limit on available operator time was 9000 hr per year.

The company used to obtain its copper wire by purchase, and the principal problem was not the planning of the extruder but the availability of the right size of wire, all of which was in short supply. Acquisition of the drawing equipment gave a new degree of flexibility. Based on anticipated prices and production and materials costs, the profit contribution of each product was determined (Table B-2a). Product details and production rates are also shown.

Burgeoning market conditions permitted sale of as much of any of the four products as the owner could make. He was not worried about having to carry some items merely to provide service and a "rounded" product line, since his industrial customers were used to shopping around and satisfying their requirements from a variety of suppliers. His question, now that he had gained new productive flexibility, was how to apportion it to the various marketable product styles so as to maximize overall profit.

Required

Utilizing a desk calculator or a computer program, ascertain the optimal program.

SOLUTION TO UNIDUCT'S PROBLEM

As shown in Table B-2b, wire products B and E are the only ones that should be manufactured, in quantities of 7.54 and 19.41 in million yards. This results in an annual profit of $15,614. The program uses all of the extruding capacity, but in drawing there is some excess of capacity over facilities employment. We have noted in the case presentation that the plastic extruder is subject to a considerable amount of downtime. A study should be made to investigate this problem further, with a view to alleviating the frequency of repairs. Cooling time might be reduced by the use of fans, which would blow air on the extruder and speed up the heat loss.

Pertinent equations are:

$$100A + 140B + 500D = 750E = Z_{max} \quad \text{(objective)}$$
$$8A + 55B + 15D = 25E = 900 \quad \text{(extruder capacity)}$$
$$60A + 40B + 400D = 400E = 9500 \quad \text{(copper drawing)}$$
$$70A + 35B + 400D = 450E = 9000 \quad \text{(labor complement)}$$

The usual matrix steps then lead to the optimal solution stated above.

Table B-2b Wire Manufacturing Programming Problem and Solution

Problem data and solution	Products saleable				Available capacity hr/yr	Total	Unused capacity hr/yr
	A	B	C	D			
a. Type conductor	Single	Single	Parallel	Parallel			
b. Gage no.	18	22	18	18			
c. No. of strands	16	7	41	41			
d. Copper no.	30	30	34	34			
e. Plastic insulation, thickness, in.	1/32	1/64	.031	.047			
f. Profit $ per million ft	100	140	500	750			
Production rate hr per million ft							
g. Plastic extruder (allowing for downtime)	8	55	15	25	900		
h. Copper drawing (3 machines)	60	40	400	400	9500		
i. Human labor (3 operators)	70	35	400	450	9000		
Optimum program							
j. Output, million ft	none	7.54	none	19.41		26.95	
k. Total profit, $/year (= $f \times j$)	none	1056	none	14558		15,614	
l. Extruding, hr/year (= $j \times g$)	none	415	none	485		900	
h. Drawing, hr/year (= $j \times h$)	none	302	none	7764		8066	
i. Labor, hr/year (= $j \times i$)	none	264	none	8736		9000	1434*

*Capacity of 9500 hr less 8060 hr used = 1434 hr unused.

CASE B-3

Evenply Wood Products Corporation: Optimum Program for Plywood Manufacturing and Marketing

Evenply currently produced six products of various quality grades, thicknesses, and plies as shown in Table B-3a. Aside from the limitations of productive capacity (shown in terms of minutes per day) in relation to attainable production rates (in minutes per 1000 sq ft), there were also market factors to be considered. As a result, no more than the following number of 1000s of square feet were considered saleable out of a day's production:

Grade 2 Plywood, 24
Stock of 3/20-in. thickness, 20
Stock of 3/8-in. thickness, 3

Table B-3a Evenply's Marketable Products and Production Rates

Problem data	A	B	C	D	E	F	Available productive capacity, min/day

Problem data	\multicolumn{6}{c}{Marketable products}	Available productive capacity, min/day
	A	B	C	D	E	F	
a. Quality grade	1	2	1	2	1	2	
b. Thickness, in.	3/20	3/20	1/4	1/4	3/8	3/8	
c. No. of plies	3	3	3	3	5	5	
d. Profit, $/1000 sq ft	10.13	15.49	1.53	7.15	5.38	10.68	
Production rates, min/1000 sq ft							
e. Veneer splicing	40.67	32.67	34.67	26.67	79.14	71.14	1920
f. Hot pressing	12.00	12.00	20.00	20.00	32.00	32.00	960
g. Sanding	20.00	20.00	20.00	20.00	20.00	20.00	960

Source: Adapted with permission from Application of Linear Programming to Plywood Production and Distribution by J. S. Bethel (Director, Wood Products Laboratory) and Cleon Harrell, Associate Professor, Economics Dept., North Carolina State College, in *Forest Products Journal*, vol. 7, no. 7, pp. 221-227, July 1957.

Considering these market limitations, the contributions to profit (and overhead) of each product, and the production rates and capacities, management desired to know the following: What products in what quantities should be made so as to take cognizance of limiting factors and yet optimize overall profit to the extent possible.

Required

1. Determine the optimum program that should serve as a guide to promotional planning and sales-production coordination.
2. Ascertain whether any production process may be creating a major bottleneck.

SOLUTION TO EVENPLY'S PROBLEM

For the purpose of best coordination of production and sales goals, the firm should produce each day the following square feet (in 1000s): 20 of A, 1 of C, 24 of D, and 3 of E. The resultant daily profit is \$391.87. Since this profit represents "contribution to profit and overhead," it may also be considered variable margin profit.

In setting up the Simplex solution steps, the objective equation becomes

$$10.13A + 15.49B \cdots\cdots + 10.68F = Z_{max}$$

Next, productive capacity is shown by

$$40.67A + 32.67B \cdots\cdots + 71.14F \leq 1920 \quad \text{(veneer splicing)}$$

$$12A + 12B \cdots\cdots + 32F \leq 960 \quad\quad\quad \text{(hot pressing)}$$

$$20A + 20B \cdots\cdots + 20F \leq 960 \quad\quad\quad \text{(sanding)}$$

Finally, market limitations are reflected in terms of additional equations:

$$B + D + F \leq 24 \quad \text{(grade 2 plywood)}$$

$$A + B \leq 20 \quad\quad \text{(3/20-in. stock)}$$

$$E + F \leq 3 \quad\quad\quad \text{(3/8-in. stock)}$$

Addition of slack variables S_1 to S_6, one to each restrictive equation, converts to equalities, and the usual Simplex procedures ensue, leading to the solution given above.

Table B-3b Plywood Programming Problem and Solution

Problem data and solution	Marketable products						Available productive capacity, min/day	Total	Unused capacity min/day
	A	B	C	D	E	F			
a. Quality grade	1	2	1	2	1	2			
b. Thickness, in.	3/20	3/20	1/4	1/4	3/8	3/8			
c. No. of plies	3	3	3	3	5	5			
d. Profit, $/1000 sq ft	10.13	15.49	1.53	7.15	5.38	10.68			
Production rates, min/1000 sq ft									
e. Veneer splicing	40.67	32.67	34.67	26.67	79.14	71.14	1920		
f. Hot pressing	12.00	12.00	20.00	20.00	32.00	32.00	960		
g. Sanding	20.00	20.00	20.00	20.00	20.00	20.00	960		
Maximum saleable, 1000 sq ft/day*									
h. Grade 2 plywood		x		x		x		24	
i. Stock of 3/20-in. thickness	x	x						20	
j. Stock of 3/8-in. thickness					x	x		3	
Optimum program									
k. Output, 1000 sq ft/day	20	none	1	24	3	none			
l. Veneer splicing, min/day (= e × k)	813	none	35	640	237	none		1725	195†
m. Hot pressing (= f × k)	240	none	20	480	96	none		836	124
n. Sanding (= g × k)	400	none	20	480	60	none		960	0
o. Profit, $/day (= d × k)	202.60	none	1.53	171.60	16.14	none		391.87	

*The x's in this section identify the products utilized for each of the three market restrictions given.

†Unused capacity represents the minutes per day that are not employed by the present program, as found from a comparison with available productive capacity (lines e, f, g). For example, for veneer splicing, 1920 minus 1725 = 195. This unused capacity can be utilized by running the present bottleneck process, sanding, overtime.

CASE B-4

Scientific Financial Trust: Portfolio Analysis

The firm's financial analyst has come up with the information below regarding three securities that are now being considered for the firm's portfolio:

Security	Cost per share, $	Expected returns, $ per share, in the next 10 years from the stocks		
		Capital growth	Cash dividends	Stock dividends
Algon Oil Corp.	5	6	1	2
Berryl Steel Co.	6	1	6	1
Duro Concrete, Inc.	8	2	4	4

The expected returns, above, are adjusted figures that allow for the effect of taxes and for present-value discounts. Thus, there need be no further concern with these details in the portfolio analysis.

It is the aim of the Trust to obtain at least the following returns for the coming 10-year period:

> Total capital growth, $6000
> Total cash dividends, $7000
> Total stock dividends, $4000

Assume that it is possible to buy fractional shares (so as to simplify the analysis).

Required

1. Determine the minimum investment required.
2. How many shares of each type (Algon, Berryl, and Duro) must be purchased?

SOLUTION TO SCIENTIFIC'S PORTFOLIO QUESTION

Because this problem consists of three variables and three restrictions per variable, it is unlikely that graphic methods will be suitable. The Simplex procedures are thus needed (Fig. B-4). From these it is apparent that an investment of $11,706 or, more precisely, $11,705.88 is required, involving 725.49 shares of Algon (A), 745.11 of Berryl (B), and 450.98 of Duro (D). The equation forms are given below:

$$5A + 6B + 9D = \min \quad \text{(objective)}$$

$$6A + B + 2D \geq 6000 \quad \text{(restriction)}$$

$$1A + 6B + 4D \geq 7000 \quad \text{(restriction)}$$

$$2F + B + 4D \geq 4000 \quad \text{(restriction)}$$

We require three slacks with coefficients − 1, also three artificial variables with a very large coefficient M. These serve to start the first matrix. The third iteration yields the solution.

Row	Coefficient	A	B	D	S_1	S_2	S_3	U_1	U_2	U_3	Result	Evaluation
U_1	$-M$	6	1	2	-1	0	0	1	0	0	6000	3000
U_2	$-M$	1	6	4	0	-1	0	0	1	0	7000	1750
U_3	$-M$	2	1	4 *(pivot)*	0	0	-1	0	0	1	4000	1000
Z		$-9M$	$-8M$	$-10M$	M	M	M	$-M$	$-M$	$-M$		
$Z\text{-}C$		$-9M+5$	$-8M+6$	$-10M+8$	M	M	M	0	0	0		
First Iteration												
U_1	$-M$	5	$\tfrac12$	0	-1	0	$\tfrac12$	1	0	$-\tfrac12$	4000	8000
U_2	$-M$	-1	5 *(pivot)*	0	0	-1	1	0	1	-1	3000	600
D	-8	$\tfrac12$	$\tfrac14$	1	0	0	$-\tfrac14$	0	0	$\tfrac14$	1000	4000
Z		$-4M-4$	$-11M/2$	-8	$+M$	$+M$	$-3M/2$	$-M$	$-M$	$3M/2$	$-7000M$ -8000	
$Z\text{-}C$		$-4M+1$	$-11M/2$	0	M	M	$-3M/2$	0	0	$5M/2$		
Second Iteration												
U_1	$-M$	5.1 *(pivot)*	0	0	-1	$1/10$	$2/5$	1	$-1/10$	$-2/5$	3700	725
B	-6	$-1/5$	1	0	0	$-1/5$	$1/5$	0	$1/5$	$-1/5$	600	-3000
D	-8	$11/20$	0	1	0	$1/20$	$-3/10$	0	$-1/20$	$3/10$	850	1545
Z		$-5.1M$	-6	-8	M	$-M/10$	$-2M/5$	$-M$	$0.1M$	$2M/5$		
Third Iteration (Final)												
A	-5	1	0	0	-5.1	$1/51$	$4/51$	5.1	$-1/51$	$-4/51$	725.49	
B	-6	0	1	0	$-2/51$	-5.1	$51/11$	$2/51$	5.1	$-11/51$	745.10	
D	-8	0	0	1	$11/102$	$2/51$	$-35/102$	$-11/102$	$-2/51$	$35/102$	450.98	
Z		-5	-6	-8	$+$	$+$	$+$	$-$	$-$	$-$	11,706	

FIG. B-4   Matrix steps leading to Scientific's solution.   Shading entries represent pivots. Some rounding has occurred in the Z and Z-C rows and in some of the column entries of the third iteration, without affecting the solution.

CASE B-5

Pepperell Manufacturing Company: Programing with Interlocking Production Alternatives and Sales Requirements

One of the mills of Pepperell Manufacturing Company is concerned with the spinning of yarn and the weaving of blankets from this yarn. Spinning involves the carding of various fiber blends, followed by a so-called "roving" operation on the fly frames and a final spinning. In each process attenuation takes place, so that a relatively coarse carded "sliver" is drafted into a finer strand known as "roving" and then further drafted and twisted into yarn. For some heavy yarns, however, no roving is required; while in other instances the fly-frame can also be used to spin even heavier yarns. In weaving, a choice of either the 92- or the 112-in. looms is available, but differences in cost result in differing profit-contributions. For example, a certain style woven on the 92-in. looms brings $101 per 100 yards, while on the 112-in. loom it brings only $98, since the wider loom must of necessity run more slowly. Production is limited by the speed of the shuttle that traverses back and forth inserting the filling, and a wider bed lengthens out this travel.

For an example, we may examine Blanket Styles no. 1 and no. 2 in Table B-5a. Although the customer sees and recognizes only *one* blanket, for programming purposes we consider this one blanket as "two styles" depending on what loom we choose to run. Only our cost and no other blanket characteristics are affected by this production choice.

The Sales Department had some special needs. It felt that two groups of Styles, nos. 4, 5, and 6 designated as Group A and 7, 8, and 9, designated as B, would have to be produced in quantities of at least 700 and 1500 yards per week as a minimum, so as to be able to offer a "balanced, full, and complete line." On the other hand, it was also felt that the market would not accept more than 7000 yards of Group A styles. These requirements are also included in Table B-5a.

Background Data

The problem data (in Table B-5a) were obtained after careful investigations. Management and the analysts realized full well that a meaningful programming solution to so complex a

Table B-5a Pepperell Manufacturing Company, Program Problem

| | Marketable blanket styles | | | | | | | | | Productive capacity per week | |
	No. 1	No. 2	No. 3	No. 4	No. 5	No. 6	No. 7	No. 8	No. 9	Machine hr	Spindle hr
Gross Profit (dollars per hundred yards sold)	101	98	60	68	65	81	55	52	58		
Time required to produce 100 yards of blankets (in hours) on:											
92-in. looms	15.8	0	17.6	18.9	0	0	15.6	0	0	2,900	
112-in. looms	0	20.2	0	0	21.6	21.6	0	17.9	17.9	900	
cards	0.30	0.30	0.36	0.36	0.36	0.45	0.31	0.31	0.34	120	
fly frames	0	0	820	830	830	950	705	705	725		125,000
spinning frames	550	550	0	0	0	0	0	0	0		20,000
Maximum volume saleable per week (100 yds) of Group A styles							}	70			
Minimum volume needed per week (100 yds) of Group A styles					} 7						
Group B styles								} 15			

Notes: 1. The maximum volume considered saleable per week, of 7000 yards for styles 7, 8, and 9 combined, is based on Sales Department experience. Similarly, a minimum of 1500 yards of these three styles is needed in order to have "a complete line" for selling purposes. Additionally, a minimum of 700 yards of Styles 4, 5, and 6 is required for "complete line" purposes.

2. Several styles, such as nos. 1 and 2, are distinct only for production purposes; viz., aside from running them on either the 92- or the 112-in. loom (at different costs and thus profits), they are carded, "fly-framed" and spun the same way and look identical as the consumer item.

Source: Kurt Eiseman and W. M. Young, Simplex Programming, a Case History, in N. L. Enrick "Management Operations Research," Chap. 8, Holt, Rinehart & Winston, New York (copyright 1965).

a problem must rest on a solid foundation of accurate and reliable figures. Readily useable data were uncovered on the speeds of machinery, productivity of processing operations, waste losses to be expected on various individual products and normal duration of downtime for maintenance and repair. The production rates derived from this information were considered relatively precise. In the separation of direct costs into fixed and variable components and in the apportionment of indirect variable costs, such as heat, light and power, it was found that many essential items were not routinely prepared. Considerable effort went into special studies to work up these data. When they had all been compiled, however, they yielded specific and reliable quantitative measures for the numerous productive operations performed in the mill. In fact, many people in management viewed this information very worthwhile by itself, quite apart from its usefulness in performing a mathematical programming analysis.

As in most programs, management took a long, hard look at fixed and variable costs with the result that only the "variable" or "out-of-pocket" costs were considered truly relevant to the programming problem. Fixed charges, it was felt, would have little relation to the selection of the proper amount of production quantities. The reason for this is that fixed items, such as depreciation on buildings, constitute long-range costs that are "sunk" and thus incurred anyway, regardless of the particular range and quantities of products that are selected for promotion in a given season or year.

It will be noted that the problem data, while realistically reflecting all of the major aspects of the facts that required consideration for programming analysis, nevertheless represent a simplification. Only a portion of the actual range of possible products and styles is shown. In this manner, needless complication is avoided, while yet preserving the essential characteristics of the real-life situation encountered. In fact, aside from the considerably larger number of products and styles, the Company also faced the problem of meeting particular customer preferences and specification for binding, stitching, "put-up" or form of folding and packaging and ticketing.

Sales forecasts provided the stipulations on volumes of production shown in Table B-5a. To assure realistic figures, the following procedure was used: Detailed preliminary forecasts were made by the sales department in the wake of a market survey, item by item. They were then revised by the manufacturing department in the light of ratios of actual orders to forecasts from past years. Furthermore, curves for long-range sales trends were reviwed. For each major item, three curves were plotted: A central curve indicating the trend (carefully prepared through statistical computations) flanked by two auxiliary curves, showing the range of fluctuation that might reasonably (at a so-called 95 percent confidence level) be expected. Any deviations of forecasts beyond these limiting curves were then ironed out by consultation between the manufacturing and sales departments.

Required

Utilizing MP analysis, determine the optimal program with regard to these factors and related requirements:

1. Quantity to be produced of each style that will maximize profit contribution.
2. Amount of available capacity utilized per week.
3. Amount idle.
4. Additional profit, in dollars per hour, obtainable if the Sales Department can see its way clear to eliminating its "balanced, full, and complete line" requirements. This additional income is known technically as the "bonus for slackening" the restrictions.

SOLUTIONS TO PEPPERELL'S PROGRAMMING PROBLEM

In seeking a solution, we set up an initial matrix as shown in Table B-5b.

Table B-5b   Pepperell Manufacturing Company, Initial Matrix of Product-mix Optimization Problem

Row	Coefficient	Body						Identity						Result R
	$C \longrightarrow$	101	98	$\cdots$ 58	0	0	0	0	$\cdots$ 0	$-M$	$-M$			
	$\downarrow$	X_1	X_2	$\cdots$ X_9	S_7	S_8	S_1	S_2	$\cdots$ S_6	U_a	U_b			
S_1	0	15.8	0	$\cdots$ 0	0	0	1	0	$\cdots$ 0	0	0			2900
S_2	0	0	20.2	$\cdots$ 17.9	0	0	0	1	$\cdots$ 0	0	0			900
S_6	0	0	0	$\cdots$ 1	0	0	0	0	$\cdots$ 1	0	0			70
U_a	$-M$	0	0	$\cdots$ 0	-1	0	0	0	$\cdots$ 0	1	0			7
U_b	$-M$	0	0	$\cdots$ 1	0	-1	0	0	$\cdots$ 0	0	1			15
				$-M$										
Z-C		-101	-98	$\cdots$ -58	0	0	0	0	$\cdots$ 0	M	M			

The dotted parts identify products X_3 to X_8 and the corresponding slack variables, S_3 to S_5. The omission of this detail from the table serves to simplify the presentation. The full initial matrix would contain a full row for each of the six slacks S_1 to S_6 and the two fictitious products U_a and U_b or a total of eight such variables in the identity portion of the matrix.

In the tableau above, X's represent blanket styles or products produced, S's represent slack variables, and U's designate artificial variables. Subscripts 1 to 9 indicate the products and the corresponding slacks, while subscripts a and b identify Groups A and B.

Successive Simplex iterations will yield the solution shown in Table B-5c.

Table B-5C Pepperell Manufacturing Company, Program Analysis

	No. 1	No. 2	No. 3	No. 4	No. 5	No. 6	No. 7	No. 8	No. 9
					Marketable blanket styles (in 100s of yards)				
Quantity to be produced	36.36	0	0	90.17	0	41.67	15.00	0	0
Penalty for violation, (in $ per 100 yards)*		5.96	7.18		6.17			5.63	1.26
Shortage from max, limit of 70:									
Group 1 †							55.0		
Excess over minimum limits of 15 and 7:									
Group A, (15) ‡							0		
Group B, (7) §				124.84					
Bonus for slackening (in $ per 100 yards)**								2.76	

Machine and spindle hours available, used and idle

	92-in. looms	112-in. looms	Cards	Fly-frame spindles	Spinning spindles
Available capacity	2900	900	120	125,000	20,000
Amount utilized	2512.77	900	66.77	125,000	20,000
Amount idle	387.23	0	52.23	0	0
Bonus for slackening (in $ per hour)**		0.15		0.08	0.18

*Penalty is the loss in total profit that occurs by producing a style that is not in the optimum, and adjusting other quantities so as to remain within all specified limits.

†The maximum salable, 70, minus the amount recommended, 15, is 55.

‡The minimum called for, 15, equals the amount recommended, 15

§The minimum called for, 7, deducted from the recommended 90.17 and 41.67, leaves a balance of 124.84.

**This represents the possible increase in overall profit when the imposed restriction is relaxed, namely, no minimum must be produced or capacity of mill is increased.

CASE B-6

Aero-Dyna Products and Engineering Corporation: Dynamic Decision Criteria

While one segment of Aero-Dyna's business is concerned with the engineering of special-purpose products and components, most of the production is devoted to relatively standardized parts for the aircraft, missile, and space industries. Only three of some 50 such items are represented in Table B-6a, followed by the solution in Table B-6b. The marketable styles are identified as E, F, and G with profit contributions based on expected prices. The two principal production processes are indicated, together with production rates for each product item and productive capacity. Market limitations, shown as "sales limit," were estimated by the sales department, based on good familiarity with the problems of supplying the various types of customers serviced by Aero-Dyna.

Table B-6a   Product Mix Problem (Lines 1–3) and Its Solution (Lines 4–5)

Description	Marketable Product Styles			Productive capacity, hr/wk
	E	F	G	
1. Variable margin profit $/unit	400	350	300	
2. Production rate, hr/unit				
Process I	350	300	400	4800
Process II	15	13	10	240
3. Sales limit, units/wk	14	10	12	
4. Optimum, units/wk	5.14	10	0	
5. Total variable-margin profit, $/wk	2056	3500	0	5556

The optimum quantity of each item, leading to an ideal profit of $5556 per week, is presented in the aforementioned solution table. This serves as a goal, guiding management action, but ignores a frequent problem of daily operations: A customer request may be received at the plant, either directly or through

Table B-6b Solution Tableau for Original Product-Mix Problem

Rows	Coefficients C	Columns						
		E	F	G	S_1	S_2	S_3	Result
		400	350	300	0	0	0	r
E	400	1	0	8/7	1/350	0	–6/7	5.14
S_0	0	0	0	–50/7	–3/70	1	–1/7	32.87
F	350	0	1	0	0	0	1	10.0
Z		400	350	457	8/7	0	8/7	5556
Z-C		0	0	157	8/7	0	8/7	

a sales representative, inquiring whether the company would accept an order that calls for weekly production of one style, say, product item G, in excess of the optimum quantity. The decision to accept or reject such an order will depend, among other thnings, upon the extent to which the resultant manufacturing commitment would unbalance capacity away from the optimal goal. Often an immediate decision must be made, so that management cannot wait until another MP analysis is performed.

It is thus necessary to develop Dynamic Decision Criteria for each product item, answering the question: If total profits are to remain above a certain level, such as 95% of the ideal attainment, then how much of the selected item (G in our example) can be produced above and beyond the original recommended amount (production of zero units in this example)? By repeating this question for each style, a set of control limits is developed *in advance* as part of the routine weekly MP analyses of the firm. These limits are then available for on–the–spot decision making, should the need arise.

Management of Aero-Dyna has decided that the optimal product quantities may be violated, so long as total profit will not decrease by more than 5% as a result of such action. In absolute terms, this percentage applied to the weekly $5556 means a reduction by $278 to $5278 of the original profit figure.

Required

Using the methods for arriving at Dynamic Decision Criteria (DDC) given in this book, develop the limits needed for each product.

SOLUTION TO AERO-DYNA PRODUCTS AND ENGINEERING CORPORATION

Although an MP solution of the optimum product mix problem has been given, a further step is required to develop DDC (Dynamic Decision Criteria), which are to be given at the 95% level. Calculation chores can be simplified by noting that

1. Sales limit of 14 units for product E exceeds the 4800 hr of capacity of Process I. A production rate of 350 hr per unit implies a requirement of 4900 hr for 14 units. Capacity is 100 hr below this, and thus the sales limit can be ignored.
2. Product F is currently recommended to be produced to the limit of attainable sales. No DDC is therefore applicable. The sales limit precludes a higher DDC value.
3. Product G is currently recommended at a level of zero output. Even after allowing some production at a 95% of optimum level, it is simply out of the question that this would exceed the current sales limit of 12 units per week.

In summary, then, the formulation of matrices leading to DDC values can safely and conveniently omit any consideration of the sales limit restrictions (three slack variables and their corresponding equations have thus been eliminated from the matrix solution problem). Moreover, no DDC for product F is required, so that DDC for only E and G will be found.

Using the previously provided solution matrix for the original problem as a starting point, we develop a new matrix (Table B-6c) which then leads to the DDC of 1.77 units for product G (Table B-6d). The plant can thus accept up to this amount of additional output on product G each week and still maintain a profit at 95% of the original optimum of $5556. Output on E is, however, reduced from 5.14 to 3.12 to accommodate this extra production on G.

For product E, the DDC is found via the Tables B-6e and f, yielding 13.71 units per week, as against the original 5.14 units. No change in G or F is required to accommodate this increase in E, but again a 5% loss in profit as against the original optimum of $5556 is experienced.

Observe that, in terms of quantity of output, E allows far greater deviations from original plans than product G. The latter may therefore be said to be more quantity-sensitive.

Table B-6c Modified Matrix to Find Dynamic Decision Criterion
for Product **G**

Rows	Coefficients C	Columns							
		E	F	G	S_1	S_2	S_3	S_4	r
		0	0	1	0	0	0	0	0
E	0	1	0	8/7	1/350	0	–6/7	0	5.14
S_2	0	0	0	–50/7	–3/70	1	–1/7	0	32.86
F	0	0	1	0	0	0	1	0	10.0
S_4	0	0	0	157	8/7	0	8/7	1	278
Z	0	0	0	0	. 0	0	0	0	0
Z-C	0	0	0	0	–1	0	0	0	0

Steps used in modifying original matrix:

1. A new slack variable S_4 has been added, with row and column as shown. The column contains zeros except 1 in the intersection with the row. The row contains the contents of the Z-C row of the original matrix to the left of this 1 and the profit reduction of 5%, applied to \$5,556, giving 278 to the right under r.
2. The coefficients C all become zero, excepting product G, which becomes 1.
3. Rows Z and Z-C are computed by the usual MP steps.
4. The modified matrix is now ready for further, conventional MP analysis to yield the DDC (see next table).

Table B-6d Solution Matrix with Dynamic Decision Criterion for Product G

Rows	Coefficients C	Columns							
		E	F	G	S_1	S_2	S_3	S_4	r
		0	0	1	0	0	0	0	0
E	0	1	0	0	–157	0	–10/11	–2/275	3.12
S_2	0	0	0	0	982	1	2/11	1/22	45.49
F	0	0	1	0	0	0	1	0	10
G	1	0	0	1	137½	0	1/22	7/1100	1.77*
Z		0	0	1	137½	0	1/22	7/1100	1.77
Z-C		0	0	0	137½	0	1/22	7/1100	1.77

*Dynamic Decision Criterion.

Table B-6e Modified Matrix to Find Dynamic Decision Criterion
for Product E

Rows	Coefficients	Columns							
		E	F	G	S_1	S_2	S_3	S_4	r
	C➤	1	0	0	0	0	0	0	0
E	1	1	0	8/7	1/350	0	–6/7	0	5.14
S_2	0	0	0	–50/7	–3/70	1	–1/7	0	32.86
F	0	0	1	0	0	0	1	0	10
S_4	0	0	0	157	8/7	0	8/7	1	278
Z		1	0	8/7	1/350	0	–6/7	0	5.14
Z-C		0	0	8/7	1/350	0	–6/7	0	5.14

Table B-6f Solution Matrix with Dynamic Decision Criterion for Product E

Rows	Coefficients	Columns							
		E	F	G	S_1	S_2	S_3	S_4	r
	C➤	1	0	0	0	0	0	0	0
E	1	1	6/7	8/7	1/350	0	0	0	13.71*
S_2	0	0	1/7	–50/7	–3/70	1	0	0	34.29
S_3	0	0	1	0	0	0	1	0	10
S_4	0	0	–50/7	157	8/7	0	0	1	206.57
Z		1	6/7	8/7	1/350	0	0	0	13.71
Z-C		0	6/7	8/7	1/350	0	0	0	13.71

*Dynamic Decision Criterion.

It is apparent that DDCs provide additional information of considerable value to the Sales Department in making decisions on both quantity and pricing.

Note: For a mathematical derivation of the procedures given here, see N. L. Enrick, Statistical Control Applications in Linear Programming, in *Management Science*, vol. 11, no. 8, pp. B–177–86, June 1965. The case data, used above, also come from this earlier source.

17

CASES SOLVABLE BY THE STEPPINGSTONE METHOD

"Steppingstone" is another word for the Distribution or Transportation method of programming. It has been chosen for the title, because we will bring case histories that involve the steppingstone method without being in the nature of requiring distribution or transportation planning as such. In other words, a more general term has been used to reflect the broad nature of applicability of the technique.

As before, problems and solution are presented separately, so that the reader's own approach to the problem is not influenced by the solution seen.

CASE C-1

Servemaster Distributors: Allocation of Vending Outlets

Servemaster had a thriving vending machine business, with three trucks supplying merchandise and collecting cash from installations in stores, service stations, and manufacturing plants. Recently problems had arisen involving excessive costs for service calls to repair inoperative machines. Management felt that the rise in repair needs was due primarily to the acquisition of the vending machines of Unicoin and Selfsell Companies, when these two firms had been bought by Servemaster. The repair service supervisor, however, thought that breakdown of equipment might be at least in part a function of location. For example, vending machines located in stores would be subject to less weather, vibrations, or dust than machines elsewhere. The experience of the past six months, in dollars of repair cost per machine, appears below:

Type of vending machine	Location of vending machine			
	Store	Service station	Light manufacturing	Heavy manufacturing
Our own, recent models	10	12	9	15
Our own, older models	8	15	12	17
Unicoin's	12	18	15	10
Selfsell's	14	9	10	18

There were 100 vending machines, 40 of Servemaster's own recent models, 10 older models, 30 Unicoin, and 20 Selfsell machines. From these the needs of stores, service stations, light manufacturing, and heavy manufacturing plants were for the following number of machines to be in place, respectively: 20, 50, 20, and 10.

Required

Analyze the problem and recommend a reallocation of machines that will be likely to minimize repair costs in the future.

SOLUTION TO SERVEMASTER'S REPAIR COST MINIMIZATION PROBLEM

Application of the distribution method of MP leads to the solution shown in Fig. C-1.

Validity of the findings depends on continuing prevalence in the future of the typical repair cost rates observed in the past for the various models and locations. Unless existing allocations are optimal, shifts in vending machines according to Fig. C-1 (or an alternate solution) are required.

It may be pointed out that factors other than repair cost may need to be considered in placing machines. For example, appearance may be a factor in sales volume. Profit per month, which takes account of sales and automatically includes repairs among the costs, would thus seem a better criterion for allocating vending machines.

The problem can rapidly assume proportions of considerable complexity when one considers that repair rates, and the waiting time from the time of breakdown until restoration to use, will also affect sales volume. Further differentiation may be required in instances where two or more machines are in one location, so that unless all of them are down simultaneously, a failure of one will merely result in an overload on the other(s).

A question can also be raised regarding the adequacy of the six-months' observation period. For example, a one-year or longer maintenance experience would be more conclusive as regards breakdown rates. Statistical methods are available to test whether or not a given period of observations is sufficient to represent practically reliable data. For the purpose of the MP analysis given here, it is assumed that management has satisfied itself in this regard.

Type of vending machine	Use-locations of vending machines				No. of machines on hand
	Store	Service station	Light manufacture	Heavy manufacture	
Our own, recent models	$10	$12 30	$9 10	$15	40
Our own, older models	$8 10	$15	$12	$17	10
Unicoin's models	$12 10	$18	$15 10	$10 10	30
Selfsell's models	$14	$9 20	$10	$18	20
Vending machines needed at locations	20	50	20	10	100*

*Current capacity equals demand.

FIG. C-1 Optimum solution to Servemaster's minimum repair cost problem. Boxed entries show repair cost per six months. Main entries show allocation of number of machines of each model to various locations. (Prepared by Milton Golenberg, student assistant.)

CASE C-2

State Finance Department: Competitive Bidding*

In order to provide for the concrete needs of four road building sites in the southeastern part of the State, the Finance Department had issued invitations to bid that resulted in three offers. Concrete producers Rockhardt, Monolith, and Brittless were able to supply weekly quantities of up to 60, 140, and 100 truckloads, respectively, at the dollar prices per cubic yard given below:

	Locations of need			
	Buddcreek	Loganscliff	Wills Hollow	Nealside
Rockhardt	12	10	11	15
Monolith	14	11	12	10
Brittless	11	9	15	8
Weekly require- ments, no. of truckloads	30	40	50	20

Price variations in these quotations were the result of different plant locations and transportation distances. Concrete quality was specified in precise terms and periodic tests were made by State highway inspectors to ensure proper proportions of constituents and strength characteristics.

Required

Determine the best allocation of contract awards to the three producers, with the aim of minimizing the Finance Department's expenditure for concrete.

*Adapted, with permission, from M. Tummins and J. F. Doom, Linear Programming, Highway Administration Applications, paper published by the Highway Research Council, Commonwealth of Virginia, Feb. 1964.

SOLUTION TO BIDDING PROBLEM

Considering the bid prices and the various needs at the locations indicated, an initial matrix is set up with a slack column, to accommodate the excess of supply offered against the amounts needed.

An optimum (Fig. C-2) is found quickly.

Bidders (plants offering concrete)	LOCATIONS OF NEED					Availability (truckloads per week)
	Buddcreek	Loganscliff	Wills Hollow	Nealside	Slack	
Rockhard	$12	$10	$11 / 50	$15	$0 / 10	60
Monolith	$14	$11	$12	$10	$0 / 140*	140
Brittless	$11 / 30	$9 / 40	$15	$8 / 20	$0 / 10	100
Requirements (truckloads per week)	30	40	50	20	160	Total 200

*In effect, Monolith's bids are all ignored, because at each location he quotes the highest prices.

FIG. C-2   Solution of bid-acceptance problem. Boxed inserts show bid prices in dollars per cubic yard of concrete. Where entries appear in a cell, they represent the number of trucks per week to be supplied. The slack column represents nonshipments of excess supply over demand.

CASE C-3

Blackwell Construction Company: Scheduling of Grading Operations*

For the construction of a highway, a 1000-foot segment of road needed leveling. Grades containing 490 cu yd of earth and low sections requiring 500 yd of fill were involved (Fig. C-3a). From a nearby borrow pit, a total of 200 yd was available for fill use. In lieu of applying the earth in the grades for filling of the low sections, it was also possible to dump as much of it as would be considered desirable.

Required

1. Develop an earth–moving program that will level the road and minimize the total transportation distance.
2. What is the minimum distance per cubic yard of earth?

---

*Adapted from same source as Case C-2.

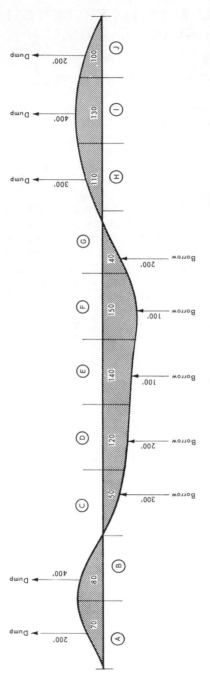
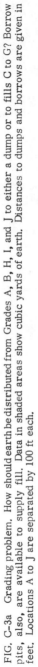

FIG. C–3a Grading problem. How should earth be distributed from Grades A, B, H, I, and J to either a dump or to fills C to G? Borrow pits, also, are available to supply fill. Data in shaded areas show cubic yards of earth. Distances to dumps and borrows are given in feet. Locations A to J are separated by 100 ft each.

SOLUTION TO BLACKWELL'S
GRADING PROBLEM

An initial matrix is set up, showing in boxed inserts the moving distances involved in each cell. From a first set of allocations, successive matrix steps will lead to a minimum total distance solution, as shown in Fig. C-3b.

The matrix distinguishes between movements that are *required*, such as to level grades and to fill low spots, and additional transportations *feasible*, such as from available borrow pit capacity to dumps. Accordingly, the matrix rim is in two parts, called "required" and "available." (A feasible move is from an available borrow or to an available dump.)

Multiplying the cubic yards of earth in each cell by the feet of travel shown in the boxed insert, and summing, gives a total of 127,000. A total of 600 cu yd of movement is shown by the cells, so that the distance per cubic yard is 211.67 ft.

A number of alternative solutions are possible, all of which give the identical end result.

In each cell below, boxed entries [ ] show distance in feet. Other (unboxed) entries show cubic yards of earth.

From \ To	C	D	E	F	G	Dumps	Total — Required to be moved	Total — Available for use
A	50 [200]	20 [300]	[400]	[500]	[600]	[200]	70	
B	[100]	80 [200]	[300]	[400]	[500]	[400]	80	
H	[500]	20 [400]	30 [300]	20 [200]	40 [100]	[300]	110	
I	[600]	[500]	[400]	130 [300]	[200]	[400]	130	
J	[700]	[600]	[500]	[400]	[300]	100 [200]	100	
Borrow pit	[300]	[200]	110 [100]	[100]	[200]	90* [0]	0	200
Total required to be moved	50	120	140	150	40	0		
Total available for use						Practically unlimited		

*This cell merely reflects the excess borrow pit capacity available. No one would actually move earth from the borrow to a dump; hence the cost is shown as zero.

FIG. C–3b Minimum–distance solution for grading project. In each cell, boxed entries show distance in feet. Other entries show cubic yards of earth.

CASE C-4

Scott Paper Company: Operations Planning from Pulp to Point of Paper Sale*

The basic objective of operations planning is to provide the customer with adequate service to keep him buying and to attain this result at the lowest overall cost to the company. Manufacturing, inventory storage, inventory handling, stock transfer costs, and transportation charges are among the primary items to be considered. Operations must be planned so that the total system costs are minimized.

Decision-making problems at Scott Paper Company can be divided into four general classifications based upon a time horizon of five years:

Type of planning	Time period
Long-range plans	5 years
Capital investment	2 years
Production scheduling	1/2 to 1 year
Warehouse operations	1 month

Long-range planning looks ahead, evaluating a large number of alternatives regarding corporate policy and strategies. Several master plans result from these analyses, providing direction and guidance for current tactical decisions, for new product priorities, for research and technology programs, and other activities. Typical projects might involve revised distribution systems, new product introductions, and modified customer service policies.

In the investment period, future problems are anticipated and resolved through acquisition of new productive capacity, enlarged distribution facilities, and related capital expenditures. The two-year span associated with this planning is based on the lead time needed to install new equipment and facilities.

Scheduling involves the utilization of available production and distribution facilities. Within the one-year period, the firm

---

*Materials due to J. R. Curtis, Director of Operations Research, Scott Paper Company, Philadelphia, Pa., as adapted from his paper, Operations Research as an Aid to Management in *Transportation and Distribution Management*, vol. 6, no. 12, pp. 23-28, Dec. 1966. Used with permission of author and publisher.

cannot invest its way out of a problem. The best it can do is to schedule existing facilities for the most efficient utilization at lowest overall costs. Operations within the final period, the current month, are necessarily quite fixed as regards handling and hauling.

Decision Processes

In order to arrive at optimal decisions, it is necessary to evaluate alternatives and arrive at a minimum cost pattern or schedule for any given arrangement of manufacturing facilities, distribution procedures, and customer service policies. For these purposes the methods of Mathematical Programming are generally suitable. The results of such analyses support the final decisions made by management.

Simplified Problem

A simplified version of MP applications utilized effectively in scheduling, investment, and planning problems at the Scott Paper Company appears in Fig. C-4a. Two paper mills supply

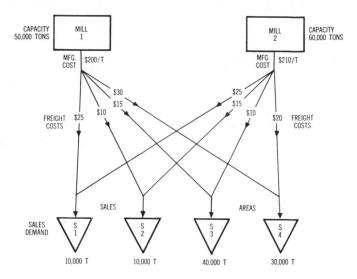

FIG. C-4a   Simplified programming problem. Two paper mills supply tonnage (T) to various sales areas (S-1 to S-4), requiring from 10,000 to 30,000 tons per month. Freight costs are indicated at $10 to 30 per ton. Manufacturing costs are $200 and 210 per ton at Warehouses 1 and 2, respectively. Given the capacities of 50,000 and 60,000 tons for the two mills, what will be the best distribution program?

a single product to four sales areas. The direct costs of production at Mill 1 are $200 per ton, while at Mill 2 the cost is $210. Freight charges vary from $10 to 30 per ton. The problem is to set up a distribution pattern whereby the total cost of meeting the demand in the system is at a minimum.

The steppingstone method will readily yield the solution. In fact the problem is small enough to permit successful application of trial and error.

Real-Life Problem

The actual situation involved a rather complex problem, which is represented in part by Fig. C-4b. Pulp can be obtained from 19 company-owned mills and from market suppliers. Some of the pulp is used in producing the 13 brands of paper, while other pulp is sold. The 8 paper mills each have warehouses that supply 20 sales areas through both solid and mixed carload shipments.

An optimal solution to this scheduling problem will require a large-scale, high-speed computer utilizing MP methodology.

Investment Decisions

A large number of investment decisions can be developed by introducing one new element into the MP method—the capital cost of expansion alternatives. Recall that the basic scheduling problem involved the use of direct costs only. These are also known as "incremental" costs, since they vary directly with increments in output. New facility alternatives must carry not only direct operating costs, but also a charge for the use of the required new capital, and thus any increase in indirect costs. Expansion evaluations are not limited to the paper mills but can also include warehouses and pulp mills. It is possible that a new production facility may have such low direct costs as to displace an older manufacturing operation in spite of the added burden of capital cost and new overhead.

Numerous subtleties of data processing and information interpretation will enter into the decision-making. Moreover, a variety of quantitative techniques from other areas of Operations Research and Management Science play significant supporting roles.

Effects of Time Periods

For long-range planning, the effects of successive time periods must be evaluated. From a viewpoint of practical

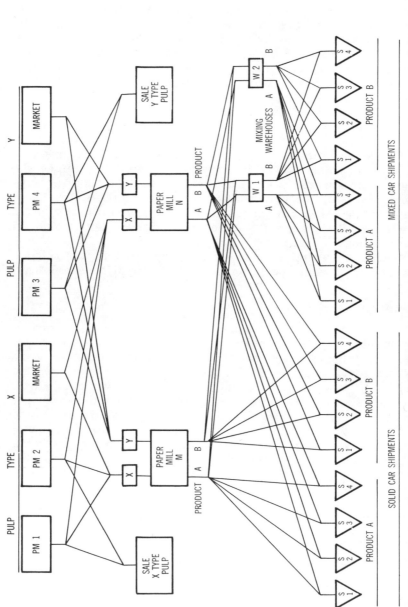

FIG. C-4b Representation of real-life problem. Pulp mills produce various types of pulp. Pulp is used by paper mills or sold to others. Pulp is also obtainable in the market. Mills ship to sales areas, either directly from their own warehouses or via mixing warehouses that facilitate mixed-car shipments.

applications, it seems best to develop an optimum arrangement of facilities, with concomitant production and distribution schedules, based on the anticipations for *each* time period. The entire system operation can then be reviewed critically as regards overall meshing, costs, and profitability. The analysis will suggest trial of alternatives, which again need evaluation. This process continues for numerous cycles, aided by computer analysis, which simulates the anticipated real situation. Once a computer has been programmed to reflect various pulp mills, paper mills, and warehouses in relation to markets and anticipated demand, it is a relatively simple matter to run repeated trials with various small alternations of arrangements per trial. From the successive computer simultations so obtained, we can further study the anticipated net cash in-flows for each period. Alternate policies and strategies can thus be tested. To allow for the effect of uncertainty and successive time periods in futurity, allowances in terms of the methods discussed in Chap. 13 on "Minimizing the Effects of Uncertainty in Planning" are made.

The system can be probed until the last increment of capital invested no longer yields an acceptable return. Such long-range planning may look tedious at first, but a few months of electronic paperwork may well be worth millions of dollars in terms of more efficient use of capital investment.

Required

Solve the simplified problem of Fig. C-4a, showing in diagram form what tonnage should go to each sales location.

SOLUTION TO SCOTT PAPER COMPANY'S PROBLEM

Either the Distribution Method or trial and error will lead to the result shown in Fig. C–4c. Although the numerical aspects of this case-problem are restricted to the simplified version of the real–life situation, nevertheless, the description of the firm's approach to various aspects of planning—with MP as a principal aid—will serve to further demonstrate how these techniques serve in the decision-making process.

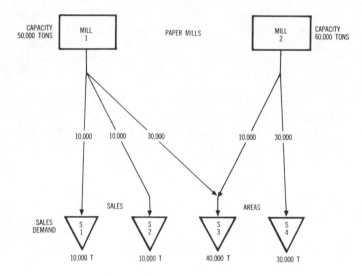

FIG. C–4c   Problem solution. The tonnage to ship from each of the two paper mills to the four sales areas shown will result in overall lowest cost distribution.

CASE C-5

Precisioneer Foundry: Materials Flow Schedules

Optimal schedules providing for the streamlined flow of materials within a plant is a further area of many useful applications of programming. Although Precisioneer's processing operations presented many opportunities for revamping of the flow of materials—raw stocks, machined items, castings, and assemblies—it was decided initially to limit investigations to the flow of skids. It seemed that some departments always experienced shortages of empty skids, while in other areas they were piling up high, causing congestion that interfered with work flow and operator safety.

A study revealed that

1. Some departments were emptying more skids than they filled. For example, the warehouse emptied the contents of skids in bins and on shelves, and later forwarded needed items via conveyor belt to shipping. Empty skids were then building up. Men were reluctant to take time in returning skids to other departments who needed them.
2. Other departments, such as casting, required a constant stream of skids to transport its output to machining and grinding. Operators would occasionally go on a search to hunt up and bring back empty skids.

Productive efficiency suffered because of lack of a definite schedule of returns. The Industrial Engineering department therefore undertook a comprehensive program of time studies and analyses, resulting in the data of Fig. C-5a. This shows

1. The number of empty skids that accumulate daily in several departments. This occurs because more skids are unloaded than filled.
2. The number of empty skids required in other departments. For example, the shapes produced in assembly are often such as not to fit snugly into tote boxes. It is therefore not unusual to find that products assembled from the contents of several tote boxes now needs additional boxes to be stored for further transport. Since one skid can safely carry only four tote boxes per layer, and a total of not more than 12 boxes, one can readily visualize the shortages that will develop in the course of a day.

Departments where more skids are filled than emptied	Departments where more skids are emptied than filled					Empty skids needed daily
	Finishing	Packing	Warehouse	Repair & scrap	Shipping	
Casting	6	1	10	5	2	15
Machining	2	4	2	7	9	9
Grinding	3	5	6	10	1	5
Stamping	2	1	9	12	5	12
Assembly	6	3	2	10	7	17
Empty skids becoming available daily	10	8	22	6	12	58

FIG. C–5a Skid return problem. In daily production, some departments use more skids than others, so that occasionally empty skids must be sent from one department to another. Transport times per empty skid are shown in small boxes (in minutes). The problem is to minimize the routing time required in returning empty skids to those departments where they are needed.

3. The time, in minutes, required to transport skids from the departments of excess accumulation to the areas of shortage.

The data are for normal operations. It was realized that further similar schedules of requirements and availabilities would have to be secured for operation at various other levels.

Required

Develop a program for the scheduled flow of skids, from areas of availability to departments of need, which will involve the minimum transport time. Based on this work, it should be possible to arrive at a sound system of materials flow that will overcome present problems.

SOLUTION TO PRECISIONEER'S MATERIALS FLOW PROBLEM

The question of optimal routing of skids from departments with excess availabilities to areas of urgent need is best approached by means of the Distribution Method of programming.

An initial route is developed in accordance with the Northwest Corner rule, as in Fig. C–5b. Successive analyses steps lead to an optimal program (Fig. C–5c). A comparison of the initial program, involving 364 min of routing time per day, as against the final 129 min, should serve to convince the skeptic that MP can be of great value in reducing costs of materials flow.

The solution found may be questioned by those who feel that transportation time will vary if several skids are returned at one time. For example, a fork–lift truck could hold six or more skids. This approach, however, is not feasible. As indicated in the case data, skids must be returned together with the empty tote boxes, thus the truck can carry only one skid per trip.

It seems desirable to lay out definite schedules, whereby specific people in each department are charged with the responsibility of piling empty skids in designated areas, so that fork truck operators will readily know where to return each skid. Once the system is in operation, a larger–scale project encompassing all of the foundry's materials flow might be undertaken. While theoretically it may appear preferable to make no changes in flows until an overall master plan has been developed, in practice the limitations of time and the need to train and orient people may well turn out to be more formidable than originally envisioned. A stepwise approach may, upon consideration of all factors, be found to be the most feasible.

Departments where more skids are filled than emptied	Departments where more skids are emptied than filled					Empty skids needed daily
	Finishing	Packing	Warehouse	Repair & scrap	Shipping	
Casting	6 ⑩ 60	1 ⑤ 5	10	5	2	15 / 65
Machining	2	4 ③ 12	2 ⑥ 12	7	9	9 / 24
Grinding	3	5	6 ⑤ 30	10	1	5 / 30
Stamping	2	1	9 ⑪ 99	12 ① 12	5	12 / 111
Assembly	6	3	2	10 ⑤ 50	7 ⑫ 84	17 / 134
Empty skids becoming available daily	10 / 60	8 / 17	22 / 141	6 / 62	12 / 84	58 / 364

FIG. C-5b Initial solution to skid return problem. The schedule developed above is based on the (previously discussed) Northwest Corner Rule. The proposed routing (such as 10 skids from Finishing to Casting, 3 from Packing to Casting, 5 from Finishing to Casting, 3 from Packing to Machining, etc.) costs 364 min of transport time. Further analysis (see next diagram) leads to an ultimate optimal schedule involving only 129 min.

Departments where more skids are filled than emptied	Departments where more skids are emptied than filled					Empty skids needed daily
	Finishing	Packing	Warehouse	Repair & scrap	Shipping	
Casting	6 ③	1	10	5 ⑤ / 25	2 ⑩ / 20	15 / 45
Machining	2 ③ / 6	4 / 10	2 ⑤	7 ① / 7	9	9 / 23
Grinding	3 ③ / 9	5 ⑧	6	10	1 ② / 2	5 / 11
Stamping	2 ④ / 8	1 / 8	9 ⑦	12	5	12 / 16
Assembly	6 / 23	3 / 34	2	10 / 32	7	17 / 34
Empty skids becoming available daily	10	8 / 44	22	6	12 / 22	58 / 129

FIG. C-5c   Solution of skid return problem. The program shown results in a minimum of 129 min per day to return skids from areas of excess accumulation to areas needing empty skids. (Prepared by Milton Golenberg, student assistant.)

Appendix

GLOSSARY OF
PROGRAMMING TERMS

Algorthmic Programming

A thorough and exhaustive mathematical approach to investigate all aspects of the given variables in a Mathematical Programming problem in order to obtain an optimal solution. Mathematical Programming is thus synonymous with algorithmic programming, but the latter term has broader implications. In particular, the term algorithmic programming is often used in contradistinction to *heuristic programming*, separately defined (see below).

Assignment Problem

Linear Programming applied to the problem of optimal assignment of resources (men, machines) among several tasks requiring completion, when each resource can be assigned to only one single functional task. The Assignment Method is often called (M. M.) Flood's method or the Hungarian (König's) method.

Convex Programming

Quadratic Programming (separately defined) in which the problem elements form what is mathematically termed a "convex set."

Discovery Method

Same as Heuristic Programming (separately defined).

Distribution Method

Same as Transportation Method.

Duality

A quality of all Linear Programming problems, referring to two ways of formulating each such problem. Thus the "dual" of a profit-maximizing problem can be given in terms of a cost-minimizing problem. The dual often gives further insights into the nature of the variables involved.

Dyadic Programming

Programming analysis to synchronize two phases of operation, each one subject to Linear Programming, with the aim of optimal meshing of plans.

Dynamic Programming

Systematic search for optimal solutions to problems that involve many highly complex interrelations that are, moreover, sensitive to multistage effects, such as successive time phases. Methods employed to solve dynamic programming problems are generally only approximately related to Mathematical Programming.

Heuristic Programming

Step-by-step search toward an optimum when a problem cannot be expressed in Mathematical Programming form. Since it is impossible to investigate all possible configurations of the variables involved, the search procedure examines successively a series of combinations that lead to step-wise improvements in the solution. The search stops when a near-optimum has been found and it becomes economically or otherwise impractical to continue further study.

Input-Output Analysis

Analysis of a complex system (organization, industry, economy) to determine the quantities of input (men, machinery, capital) required to attain certain outputs (goods and services). Input-output analysis is considered a Mathematical Programming problem, but the method of finding solutions is somewhat different.

Integer Programming

Linear Programming in which the solution is required in terms of integers (whole numbers).

Linear Programming

Mathematical Programming in which all variables are stated in terms of linear relationships. This is the most frequent form of programming problems.

Mathematical Programming

A systematic method for analyzing the relationships among several interdependent variables, with the objective of ascertaining a maximal or minimum point as the solution. Usually there is a goal or objective equation (separately defined) in terms of gains or costs that need maximizing or minimizing, while other equations state the limiting factors that act as constraints.

Matrix

Same as Tableau (separately defined).

Matrix Theory

A branch of algebra on which the procedures of Mathematical Programming are based.

Modi Method

A modification of the Distribution Method of Linear Programming.

Nonlinear Programming

Mathematical Programming in which some or all of the variables are curvilinear. Many types of nonlinear forms cannot be fully solved by presently known methods.

Parametric Programming

Linear Programming modified for the purpose of inclusion of several objective equations with varying degrees of priority. The sensitivity of the solution to these variations is then studied. See also Sensitivity Analysis.

PERT—Program Evalution and Review Technique

Method for analysis of the network of activities in large-scale projects involving complex and interlocking sequences of operation. A critical path is determined, representing the sum of individual time values in a particular sequence of activities that has no leeway or "slack" in it. Therefore, if the time taken to perform any operation in the critical path increases, total project time will be lengthened by that amount. Expected project completion dates, materials requirements, equipment and manpower needs, and costs can thus be evaluated and brought under supervision in a systematic manner. Many of the problem aspects of PERT are similar to Mathematical Programming, but the approach to a solution is distinct.

Probabilistic Programming

Linear Programming that includes an evaluation of relative risks, and uncertainties in various alternatives of choice for management decisions. These uncertainties may involve considerations of whether or not a certain sales volume can be attained or the likelihood of persistence over a period of future time of certain market factors.

Quadratic Programming

A modification of Linear Programming, in which the objective equations appear in quadratic form, that is, they contain squared terms.

Ratio Analysis Programming

An approximate method for evaluation of Mathematical Programming problems, without seeking a definitive optimum.

Sensitivity Analysis

Analysis of the sensitivity or degree of response of a Linear Programming solution to varying degrees of change in the problem variables. See also Parametric Programming.

Simplex Programming

Same as Linear Programming. The word "simplex" does not imply simplicity, but refers to certain details of matrix algebra (separately defined) used in formulating the problem that is solved by Linear Programming.

Stochastic Programming

Same as Probabilistic Programming.

Tableau

The particular form in which a system of equations, containing the Mathematical Programming problem and, eventually, its solution, is expressed in a table. This form is maintained through a series of so-called recursive or iterative steps until the final Tableau yields the problem solution. In lieu of Tableau, the more general term "matrix" is often used. Both terms refer to a rectangular array of data.

Transportation Method

A simplified form of Linear Programming. It is applicable only in those instances where goals and constraints imposed by limited resources are expressed in identical units and, moreover, there are no limiting interdependencies among the constraints.

VAM Programming

A short-cut method pertaining to the Transportation Programming procedure. VAM is an abbreviation for (William R.) Vogel's Approximation Method.

INDEX